1984 YEARBOOK of ASTRONOMY

EDITED BY PATRICK MOORE

1984 YEARBOOK of ASTRONOMY

EDITED BY PATRICK MOORE

W · W · Norton & Company
NEW YORK LONDON

First American Edition 1984

ISBN 0-393-30147-8

Printed in Great Britain

Contents

PART TWO: ARTICLE SECTION

PART THREE: MISCELLANEOUS

Editor's Foreword

The pattern of the *1984 Yearbook* follows the now-familiar line. As usual, the data for 1984 have been compiled by Gordon Taylor of the Royal Greenwich Observatory; at the suggestion of Sir Dennis Proctor, extra lunar data have been added. The Article section covers a wide variety of topics, and as well as our regular and valued contributors, notably Dr David Allen, we have some new contributors of great distinction. We have again done our best to provide 'something for everybody'. My thanks are due to Barney D'Abbs for invaluable help in proof-reading.

PATRICK MOORE

Selsey, May 1983

Preface

New readers will find that all the information in this *Yearbook* is given in diagrammatic or descriptive form; the positions of the planets may easily be found on the specially designed star charts, while the monthly notes describe the movements of the planets and give details of other astronomical phenomena visible in both the northern and southern hemispheres. Two sets of star charts are provided. The **Northern Charts** (pp. 16 to 41) are designed for use in latitude 52 degrees north, but may be used without alteration throughout the British Isles, and (except in the case of eclipses and occultations) in other countries of similar north latitude. The **Southern Charts** (pp. 42 to 67) are drawn for latitude 35 degrees south, and are suitable for use in South Africa, Australia and New Zealand, and other stations in approximately the same south latitude. The reader who needs more detailed information will find *Norton's Star Atlas* (Gall and Inglis) an invaluable guide, while more precise positions of the planets and their satellites, together with predictions of occultations, meteor showers, and periodic comets may be found in the *Handbook* of the British Astronomical Association. A somewhat similar publication is the *Observer's Handbook* of the Royal Astronomical Society of Canada, and readers will also find details of forthcoming events given in the American *Sky and Telescope*. This monthly publication also produces a special occultation supplement giving predictions for the United States and Canada.

Important Note

The times given on the star charts and in the Monthly Notes are generally given as local times, using the 24-hour clock, the day beginning at midnight. All the dates, and the times of a few events (e.g. eclipses), are given in Greenwich Mean Time (G.M.T.), which is related to local time by the formula

Local Mean Time = G.M.T. – west longitude.

In practice, small differences of longitude are ignored, and the observer will use local clock time, which will be the appropriate Standard (or Zone) Time. As the formula indicates, places in west longitude will have a standard Time slow on G.M.T., while places in east longitude will have Standard Times fast on G.M.T. As examples we have:

Standard Time in

New Zealand	G.M.T.	+	12 hours
Victoria; N.S.W.	G.M.T.	+	10 hours
Western Australia	G.M.T.	+	8 hours
South Africa	G.M.T.	+	2 hours
British Isles	G.M.T.		
Eastern S.T.	G.M.T.	–	5 hours
Central S.T.	G.M.T.	–	6 hours, etc.

If Summer Time is in use, the clocks will have been advanced by one hour, and this hour must be subtracted from the clock time to give Standard Time.

In Great Britain and N. Ireland, Summer Time will be in force in 1984 from March $25^d 01^h$ to October $28^d 01^h$.

PART ONE

Monthly Charts and Astronomical Phenomena

Notes on the Star Charts

The stars, together with the Sun, Moon and planets seem to be set on the surface of the celestial sphere, which appears to rotate about the Earth from east to west. Since it is impossible to represent a curved surface accurately on a plane, any kind of star map is bound to contain some form of distortion. But it is well known that the eye can endure some kinds of distortion better than others, and it is particularly true that the eye is most sensitive to deviations from the vertical and horizontal. For this reason the star charts given in this volume have been designed to give a true representation of vertical and horizontal lines, whatever may be the resulting distortion in the shape of a constellation figure. It will be found that the amount of distortion is, in general, quite small, and is only obvious in the case of large constellations such as Leo and Pegasus, when these appear at the top of the charts, and so are drawn out sideways.

The charts show all stars down to the fourth magnitude, together with a number of fainter stars which are necessary to define the shape of a constellation. There is no standard system for representing the outlines of the constellations, and triangles and other simple figures have been used to give outlines which are easy to follow with the naked eye. The names of the constellations are given, together with the proper names of the brighter stars. The apparent magnitudes of the stars are indicated roughly by using four different sizes of dots, the larger dots representing the bright stars.

The two sets of star charts are similar in design. At each opening there is a group of four charts which give a complete coverage of the sky up to an altitude of 62½ degrees; there are twelve such groups to cover the entire year. In the **Northern Charts** (for 52 degrees north) the upper two charts show the southern sky, south being at the centre and east on the left. The coverage is from 10 degrees north of east (top left) to 10 degrees north of west (top right). The two lower charts show the northern sky from 10 degrees south of west (lower left) to 10 degrees south of east (lower right). There is thus an overlap east and west.

Conversely, in the **Southern Charts** (for 35 degrees south) the upper two charts show the northern sky, with north at the centre and east on the right. The two lower charts show the southern sky, with south at the centre and east on the left. The coverage and overlap is the same on both sets of charts.

Because the sidereal day is shorter than the solar day, the stars appear to rise and set about four minutes earlier each day, and this amounts to two hours in a month. Hence the twelve groups of charts in each set are sufficient to give the appearance of the sky throughout the day at intervals of two hours, or at the same time of night at monthly intervals throughout the year. The actual range of dates and times when the stars on the charts are visible is indicated at the top of each page. Each group is numbered in bold type, and the number to be used for any given month and time is summarized in the following table:

Local Time	18h	20h	22h	0h	2h	4h	6h
January	11	12	1	2	3	4	5
February	12	1	2	3	4	5	6
March	1	2	3	4	5	6	7
April	2	3	4	5	6	7	8
May	3	4	5	6	7	8	9
June	4	5	6	7	8	9	10
July	5	6	7	8	9	10	11
August	6	7	8	9	10	11	12
September	7	8	9	10	11	12	1
October	8	9	10	11	12	1	2
November	9	10	11	12	1	2	3
December	10	11	12	1	2	3	4

The charts are drawn to scale, the horizontal measurements, marked at every 10 degrees, giving the azimuths (or true bearings) measured from the north round through east (90 degrees), south (180 degrees), and west (270 degrees). The vertical measurements, similarly marked, give the altitudes of the stars up to 62½ degrees. Estimates of altitude and azimuth made from these charts will necessarily be mere approximations, since no observer will be exactly at the adopted latitude, or at the stated time, but they will serve for the identification of stars and planets.

The ecliptic is drawn as a broken line on which longitude is marked at every 10 degrees; the positions of the planets are then easily found by reference to the table on page 74. It will be noticed that on the southern charts the ecliptic may reach an altitude in excess of 62½ degrees on star charts 5 to 9. The continuation of the broken line will be found on the charts of overhead stars.

There is a curious illusion that stars at an altitude of 60 degrees or more are actually overhead, and the beginner may often feel that he is leaning over backwards in trying to see them. These overhead stars are given separately on the pages immediately following the main star charts. The entire year is covered at one opening, each of the four maps showing the overhead stars at times which correspond to those of three of the main star charts. The position of the zenith is indicated by a cross, and this cross marks the centre of a circle which is 35 degrees from the zenith; there is thus a small overlap with the main charts.

The broken line leading from the north (on the Northern Charts) or from the south (on the Southern Charts) is numbered to indicate the corresponding main chart. Thus on page 40 the N-S line numbered 6 is to be regarded as an extension of the centre (south) line of chart 6 on pages 26 and 27, and at the top of these pages are printed the dates and times which are appropriate. Similarly, on page 67, the S-N line numbered 10 connects with the north line of the upper charts on pages 60 and 61.

The overhead stars are plotted as maps on a conical projection, and the scale is rather smaller than that of the main charts.

1L

October 6 at 5ʰ	October 21 at 4ʰ
November 6 at 3ʰ	November 21 at 2ʰ
December 6 at 1ʰ	December 21 at midnight
January 6 at 23ʰ	January 21 at 22ʰ
February 6 at 21ʰ	February 21 at 20ʰ

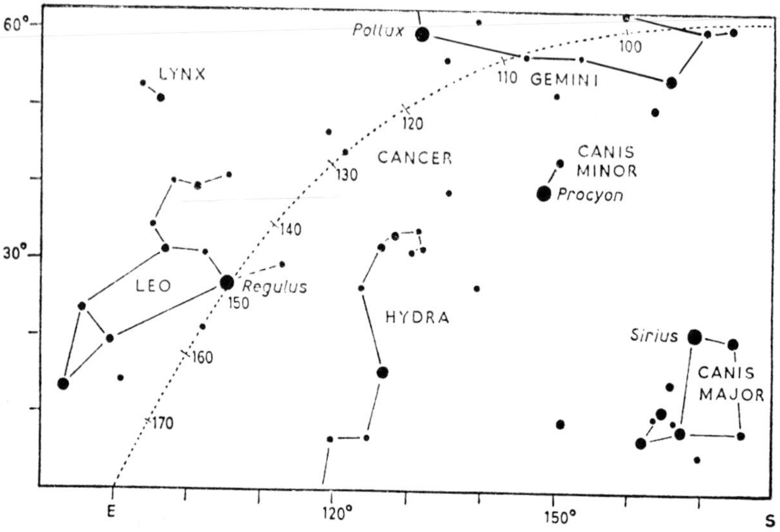

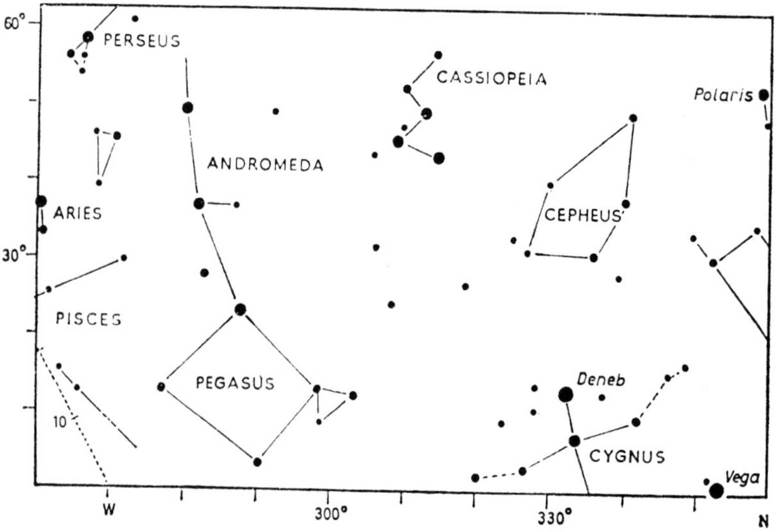

October 6 at 5h	October 21 at 4h	
November 6 at 3h	November 21 at 2h	
December 6 at 1h	December 21 at midnight	**1R**
January 6 at 23h	January 21 at 22h	
February 6 at 21h	February 21 at 20h	

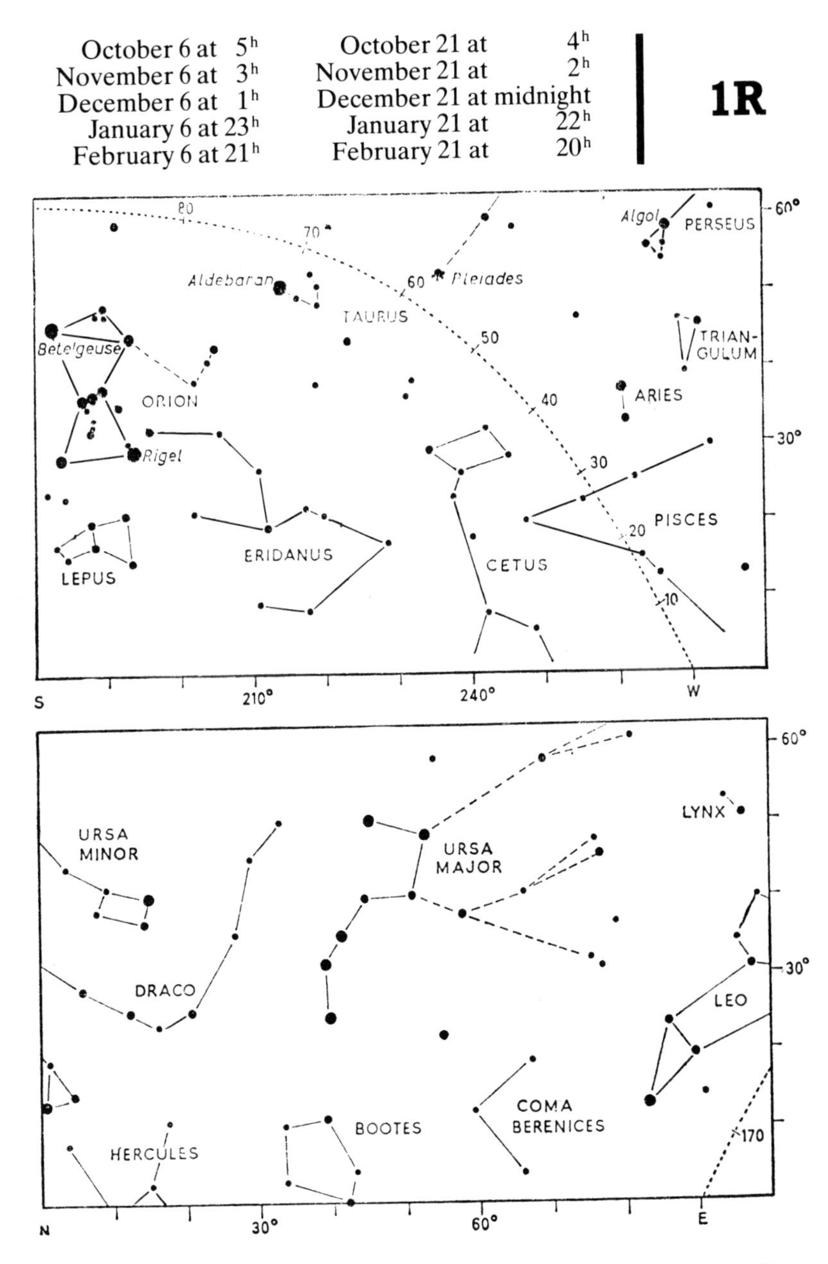

2L

November 6 at 5^h	November 21 at 4^h
December 6 at 3^h	December 21 at 2^h
January 6 at 1^h	January 21 at midnight
February 6 at 23^h	February 21 at 22^h
March 6 at 21^h	March 21 at 20^h

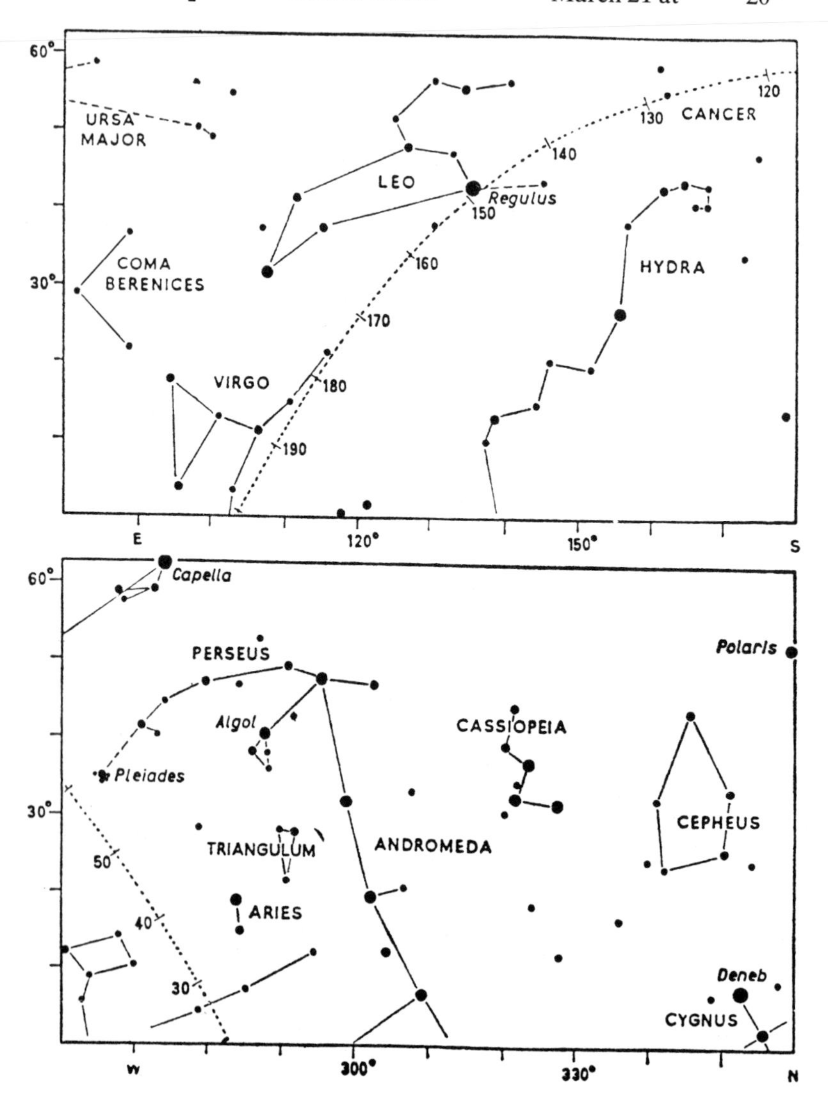

November 6 at 5^h	November 21 at 4^h
December 6 at 3^h	December 21 at 2^h
January 6 at 1^h	January 21 at midnight
February 6 at 23^h	February 21 at 22^h
March 6 at 21^h	March 21 at 20^h

2R

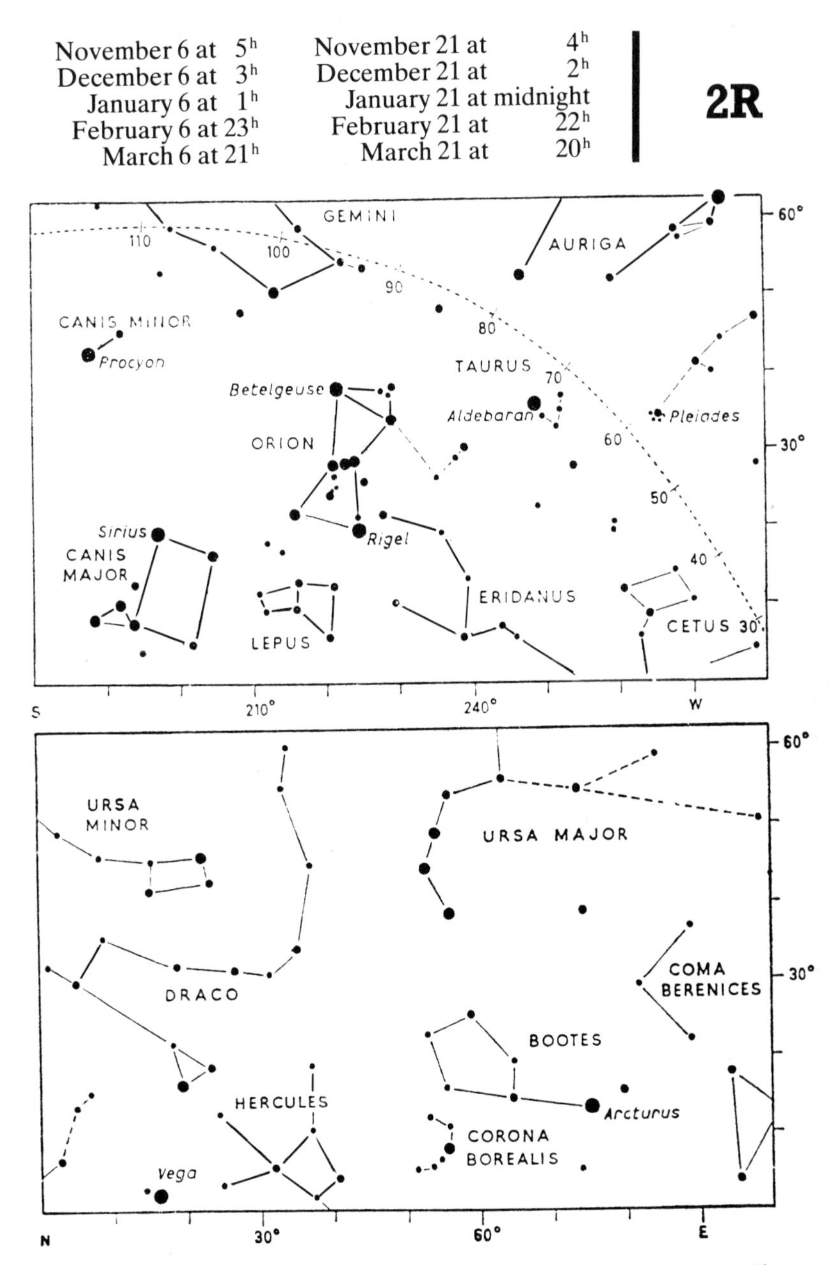

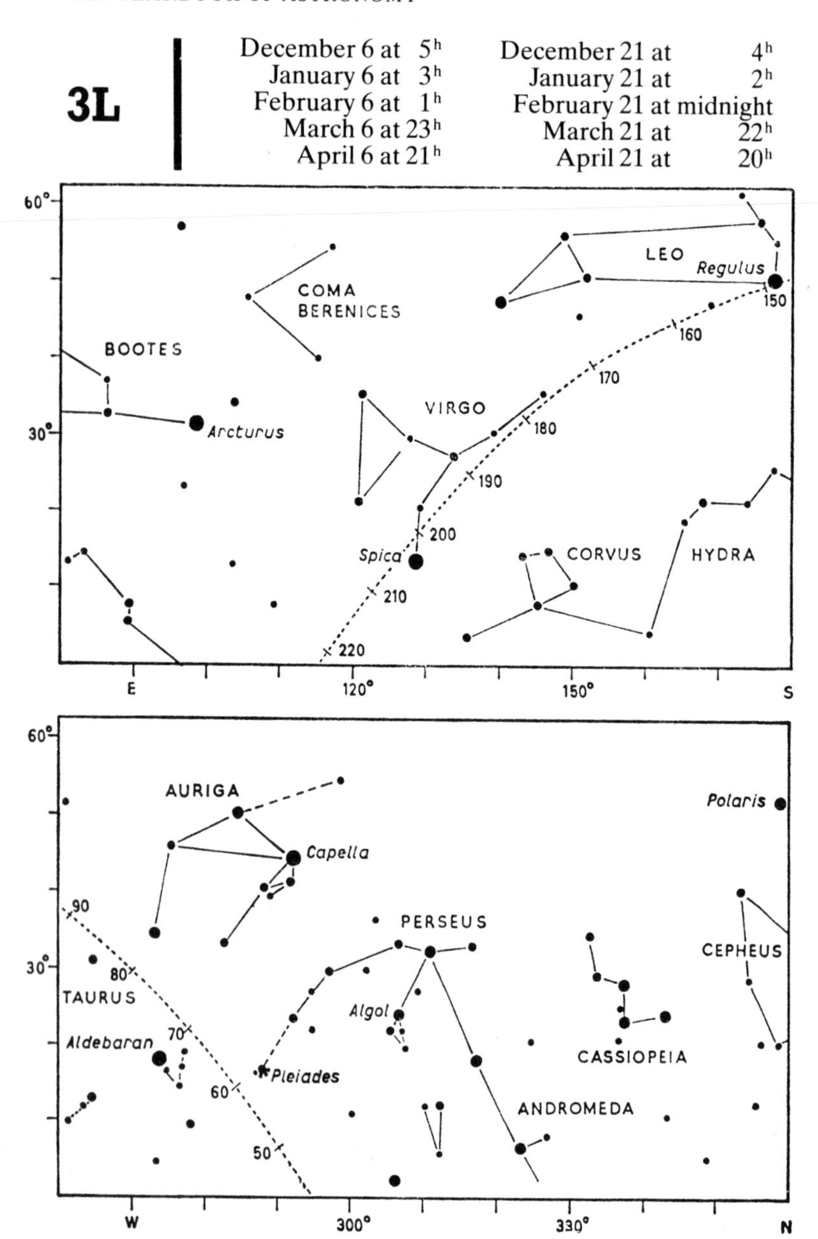
3L
December 6 at 5h
January 6 at 3h
February 6 at 1h
March 6 at 23h
April 6 at 21h
December 21 at 4h
January 21 at 2h
February 21 at midnight
March 21 at 22h
April 21 at 20h
60°
30°
LEO
Regulus
COMA BERENICES
BOOTES
Arcturus
VIRGO
Spica
CORVUS
HYDRA
150
160
170
180
190
200
210
220
E
120°
150°
S
AURIGA
Capella
Polaris
PERSEUS
CEPHEUS
TAURUS
Algol
Aldebaran
CASSIOPEIA
Pleiades
ANDROMEDA
90
80
70
60
50
W
300°
330°
N

December 6 at 5^h	December 21 at 4^h
January 6 at 3^h	January 21 at 2^h
February 6 at 1^h	February 21 at midnight
March 6 at 23^h	March 21 at 22^h
April 6 at 21^h	April 21 at 20^h

3R

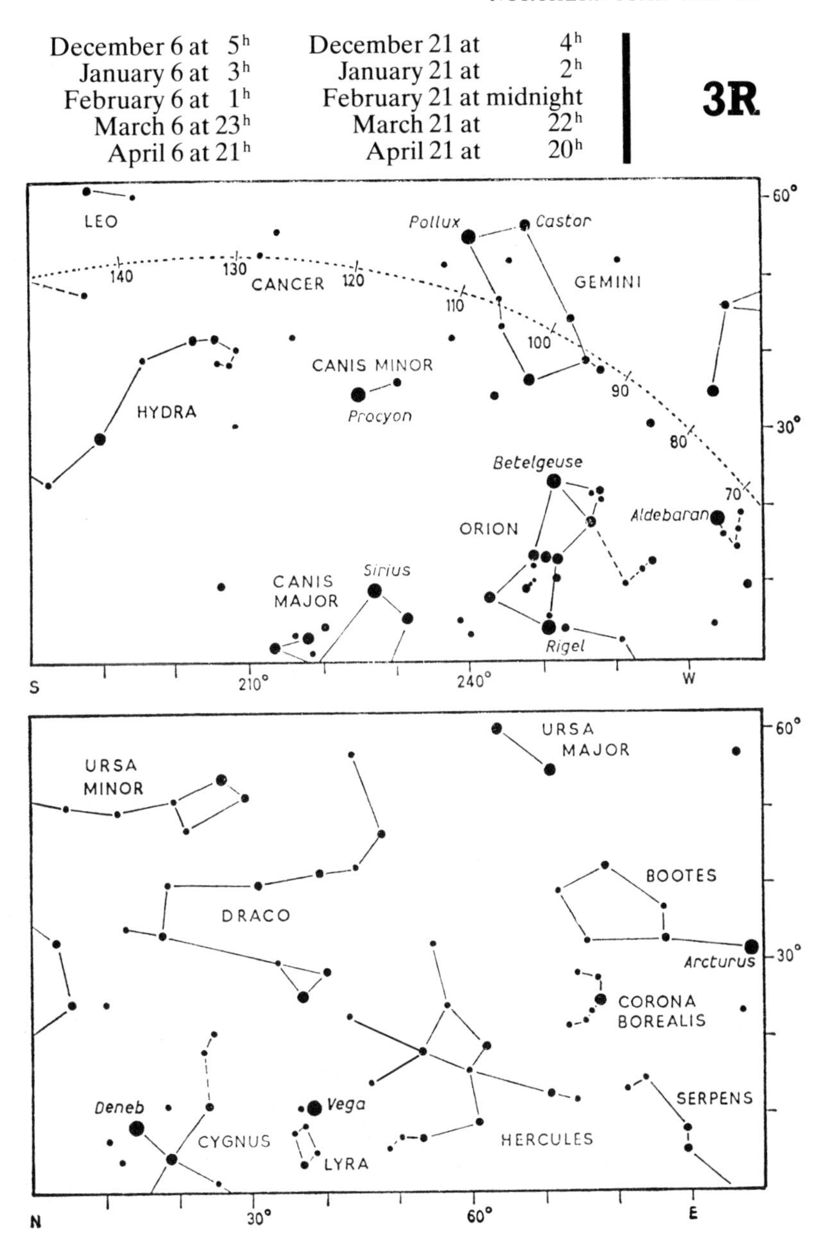

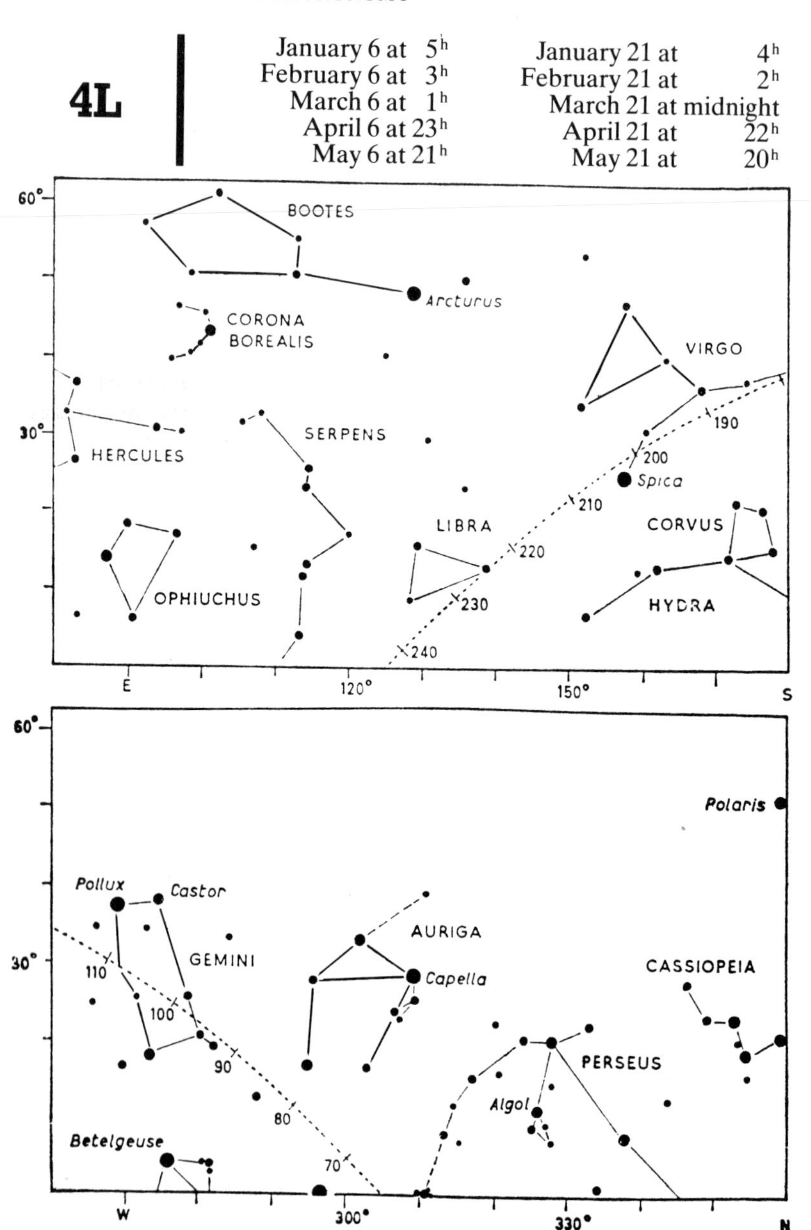
4L
January 6 at 5h
February 6 at 3h
March 6 at 1h
April 6 at 23h
May 6 at 21h
January 21 at 4h
February 21 at 2h
March 21 at midnight
April 21 at 22h
May 21 at 20h
60°
30°
BOOTES
Arcturus
CORONA BOREALIS
VIRGO
HERCULES
SERPENS
190
200
Spica
210
LIBRA
CORVUS
220
OPHIUCHUS
230
HYDRA
240
E
120°
150°
S
Polaris
Pollux
Castor
GEMINI
AURIGA
Capella
CASSIOPEIA
110
100
90
PERSEUS
Algol
80
Betelgeuse
70
W
300°
330°
N

January 6 at 5h	January 21 at 4h	**4R**
February 6 at 3h	February 21 at 2h	
March 6 at 1h	March 21 at midnight	
April 6 at 23h	April 21 at 22h	
May 6 at 21h	May 21 at 20h	

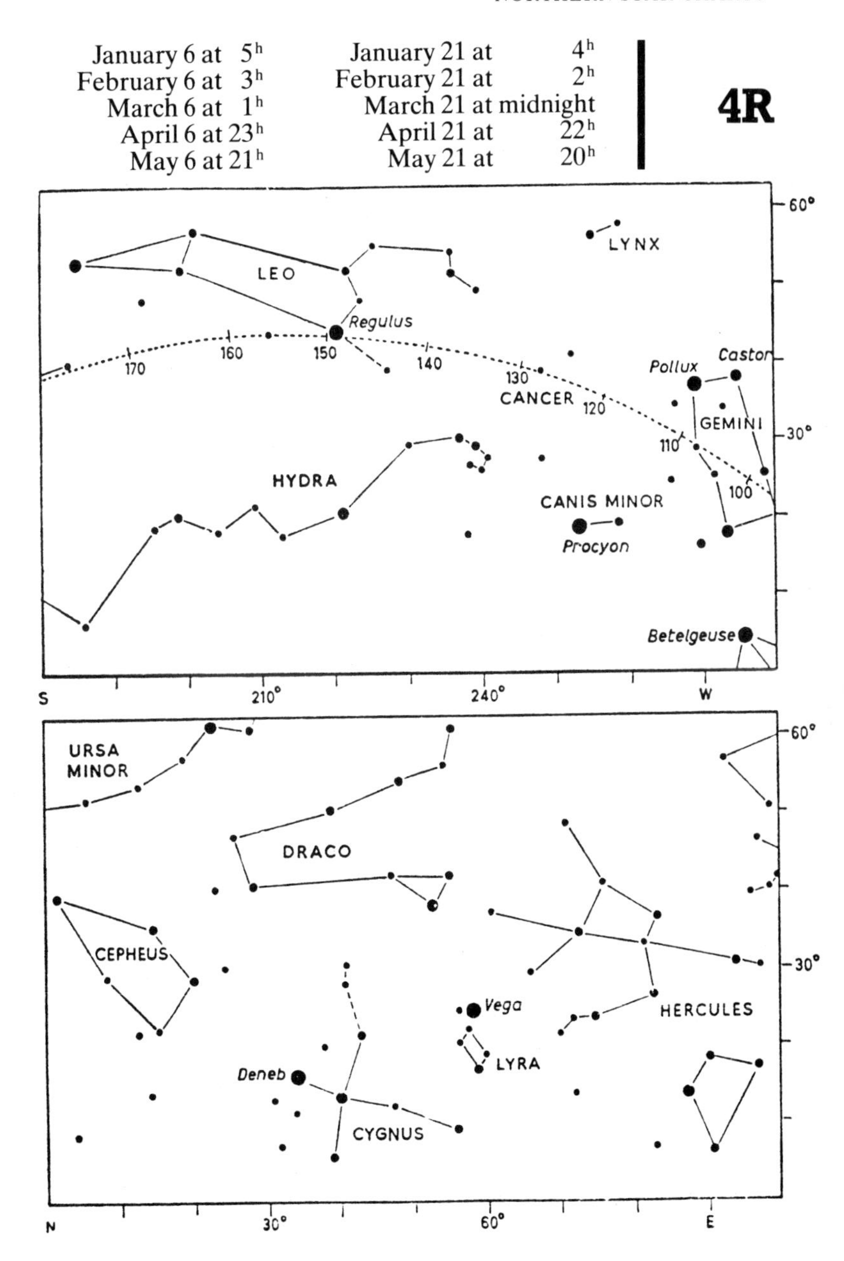

5L

January 6 at 7ʰ	January 21 at 6ʰ
February 6 at 5ʰ	February 21 at 4ʰ
March 6 at 3ʰ	March 21 at 2ʰ
April 6 at 1ʰ	April 21 at midnight
May 6 at 23ʰ	May 21 at 22ʰ

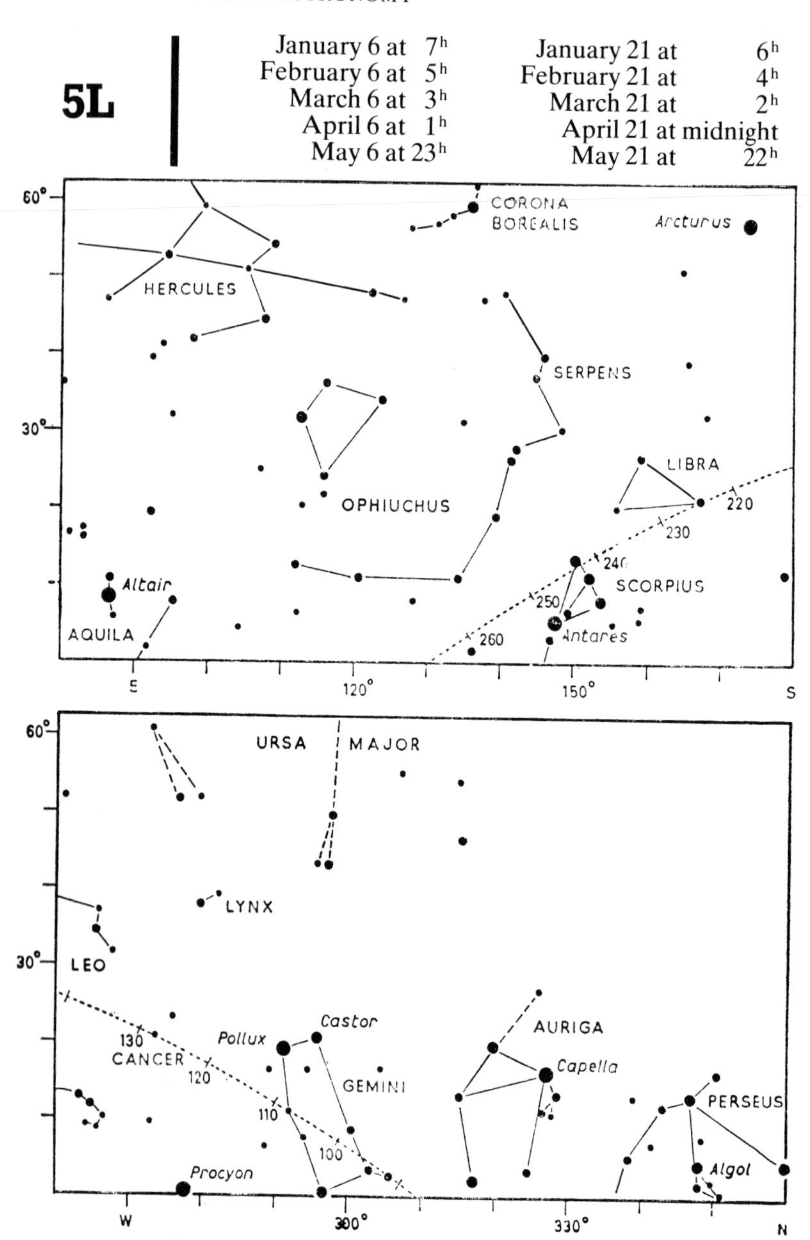

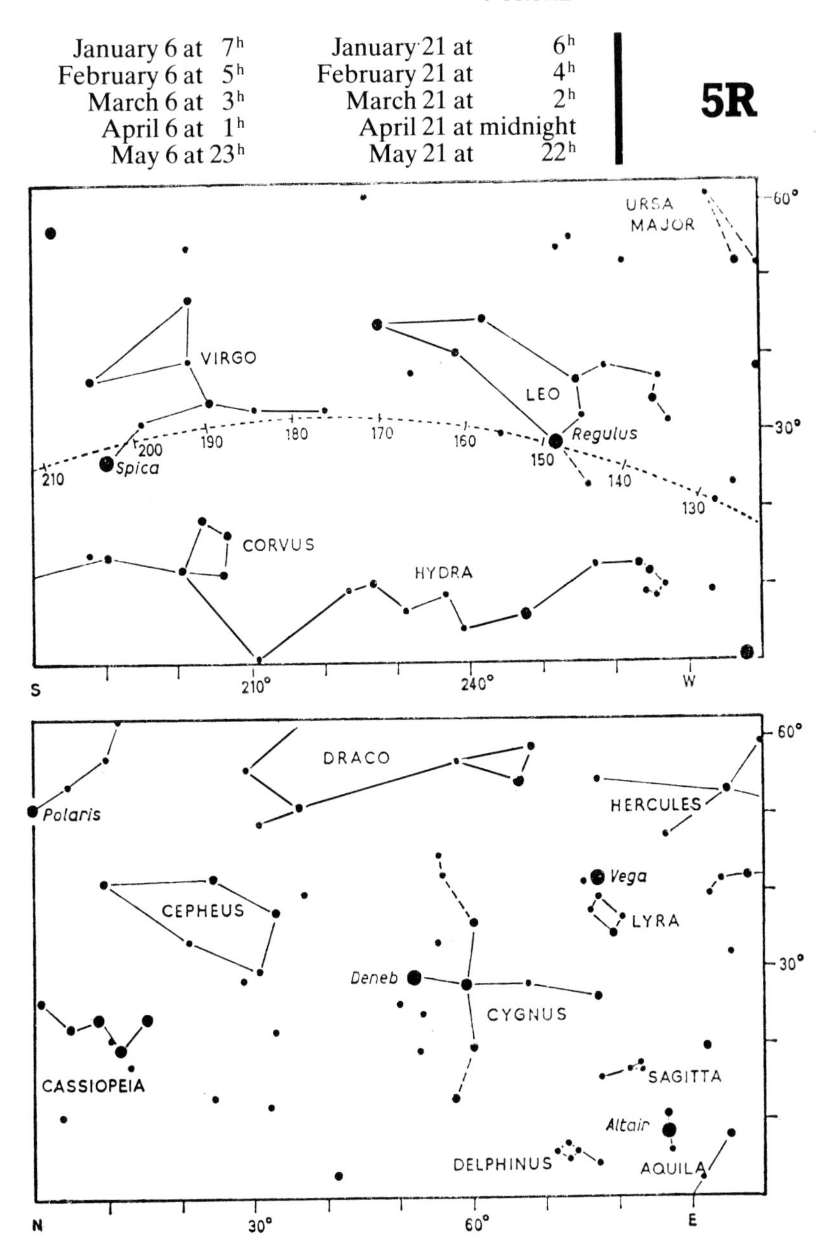

January 6 at 7h
February 6 at 5h
March 6 at 3h
April 6 at 1h
May 6 at 23h
January 21 at 6h
February 21 at 4h
March 21 at 2h
April 21 at midnight
May 21 at 22h
5R
URSA MAJOR
VIRGO
LEO
Regulus
Spica
CORVUS
HYDRA
210
200
190
180
170
160
150
140
130
60°
30°
S
210°
240°
W
DRACO
HERCULES
Polaris
Vega
LYRA
CEPHEUS
Deneb
CYGNUS
CASSIOPEIA
SAGITTA
Altair
DELPHINUS
AQUILA
N
30°
60°
E

6L

March 6 at 5h	March 21 at 4h
April 6 at 3h	April 21 at 2h
May 6 at 1h	May 21 at midnight
June 6 at 23h	June 21 at 22h
July 6 at 21h	July 21 at 20h

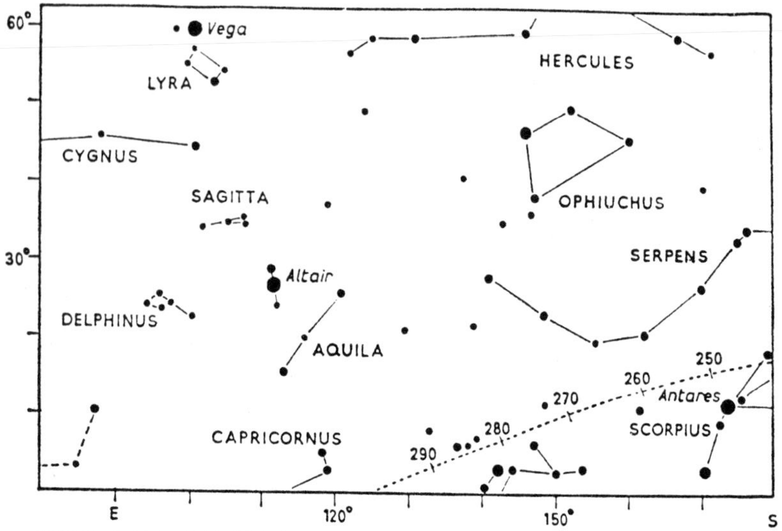

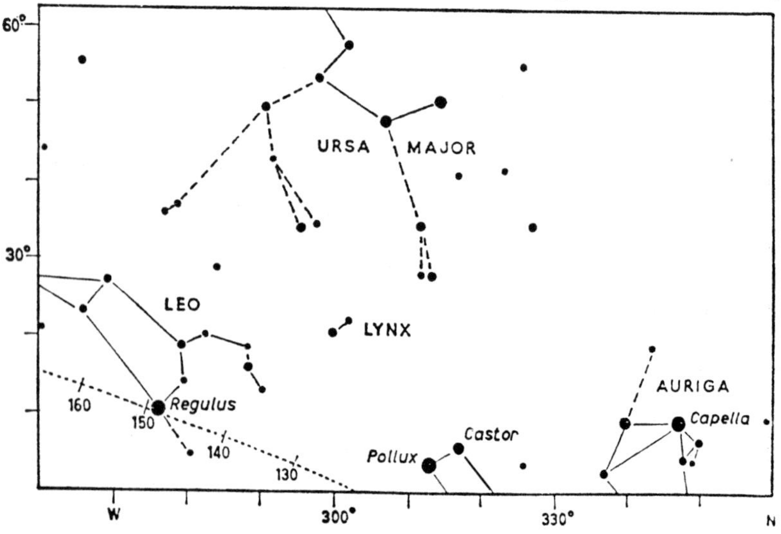

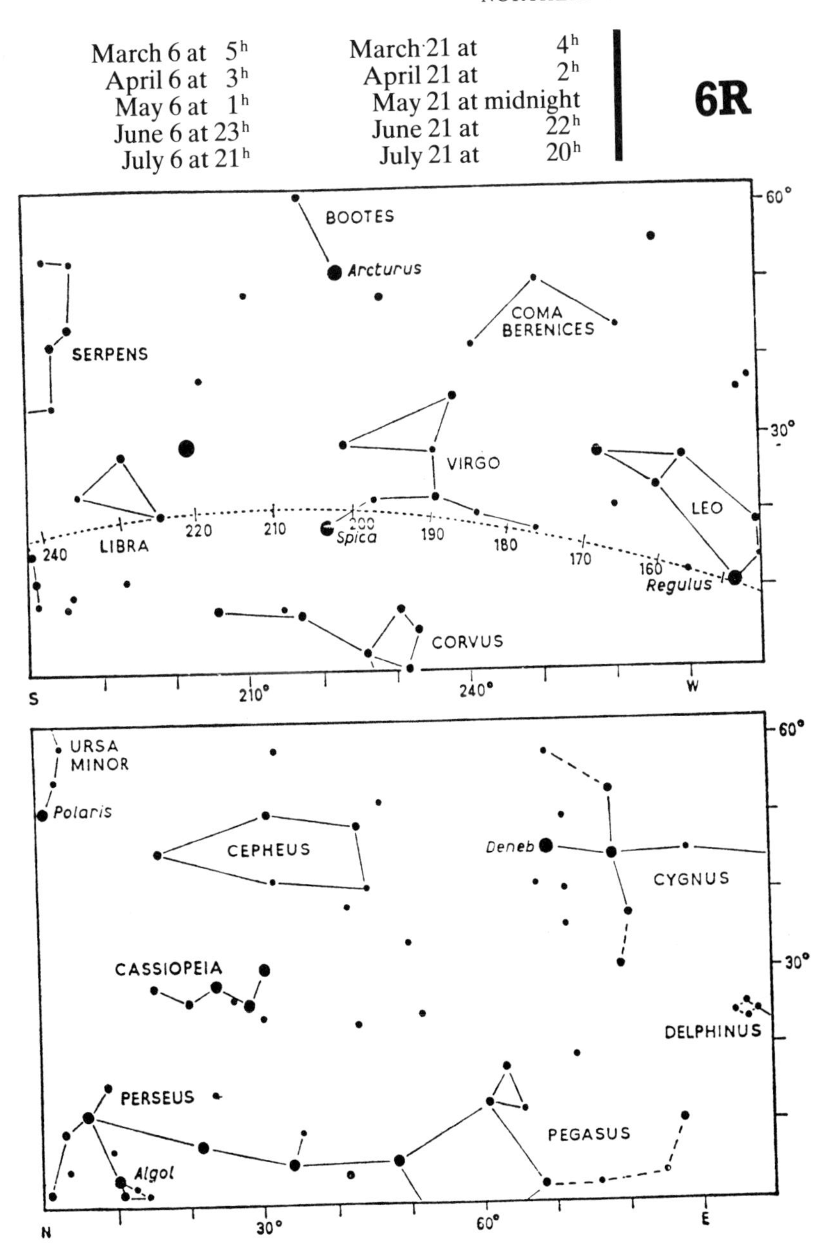
March 6 at 5h
April 6 at 3h
May 6 at 1h
June 6 at 23h
July 6 at 21h
March 21 at 4h
April 21 at 2h
May 21 at midnight
June 21 at 22h
July 21 at 20h
6R
BOOTES
Arcturus
COMA BERENICES
SERPENS
VIRGO
LEO
LIBRA
Spica
Regulus
CORVUS
240
220
210
200
190
180
170
160
60°
30°
S
210°
240°
W
URSA MINOR
Polaris
CEPHEUS
Deneb
CYGNUS
CASSIOPEIA
DELPHINUS
PERSEUS
Algol
PEGASUS
60°
30°
N
30°
60°
E

7L

May 6 at 3h	May 21 at 2h
June 6 at 1h	June 21 at midnight
July 6 at 23h	July 21 at 22h
August 6 at 21h	August 21 at 20h
September 6 at 19h	September 21 at 18h

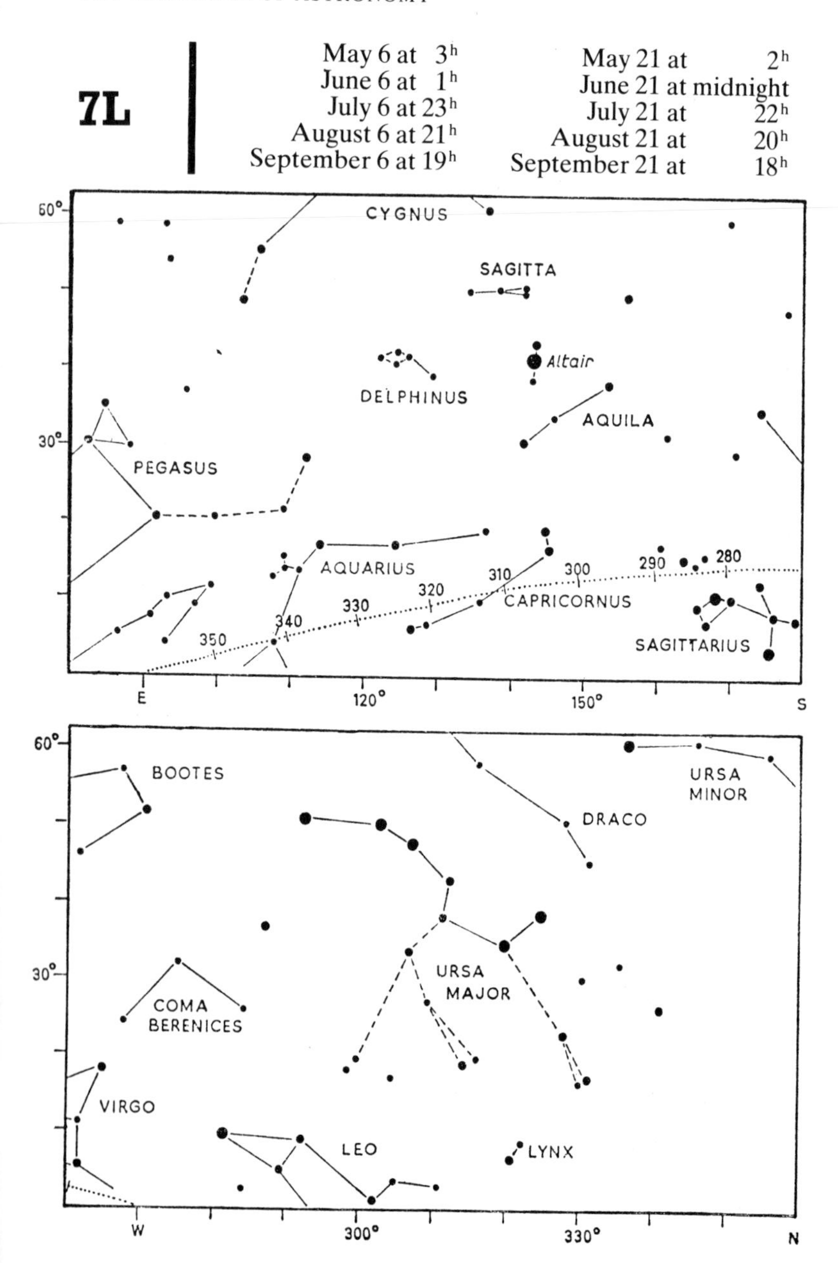

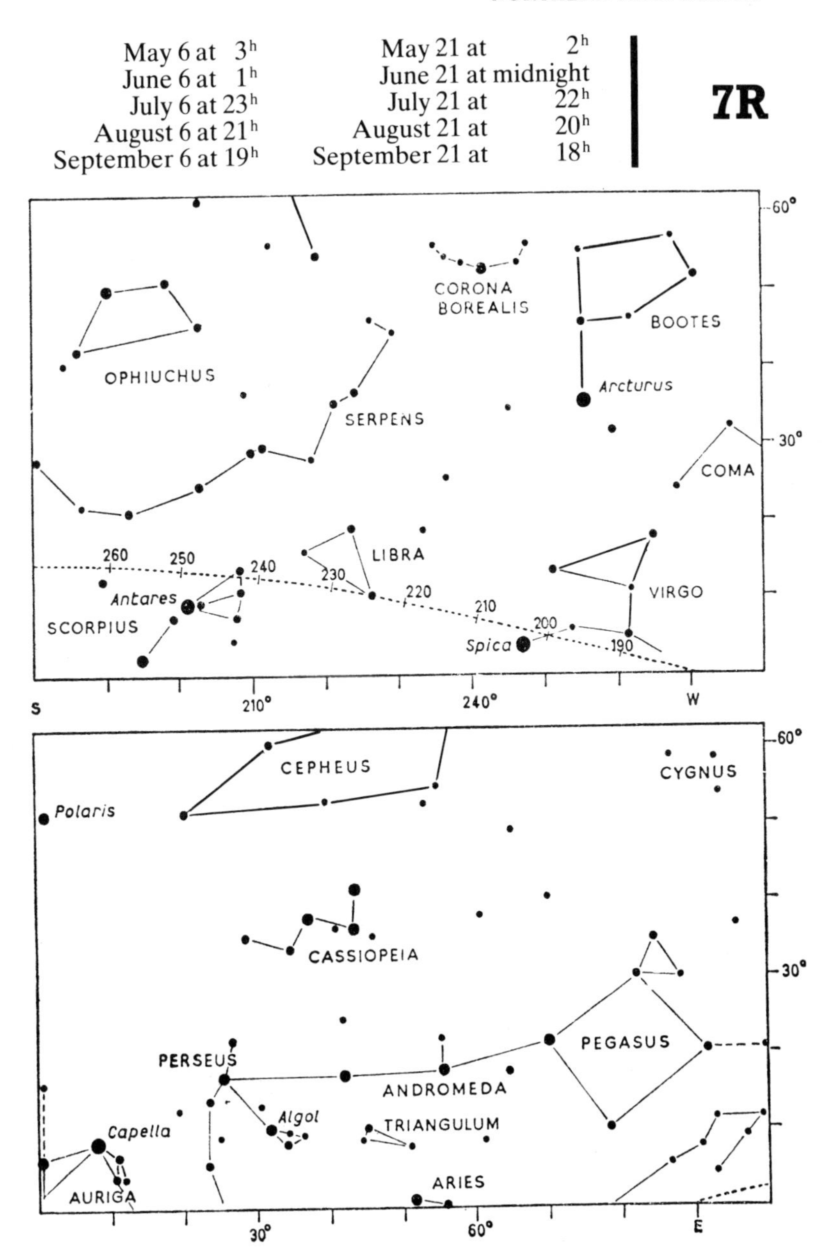
May 6 at 3^h
June 6 at 1^h
July 6 at 23^h
August 6 at 21^h
September 6 at 19^h
May 21 at 2^h
June 21 at midnight
July 21 at 22^h
August 21 at 20^h
September 21 at 18^h
7R
CORONA BOREALIS
BOOTES
OPHIUCHUS
Arcturus
SERPENS
COMA
LIBRA
VIRGO
260
250
240
230
220
210
200
190
Antares
SCORPIUS
Spica
60°
30°
S
210°
240°
W
CEPHEUS
CYGNUS
Polaris
CASSIOPEIA
PEGASUS
PERSEUS
ANDROMEDA
Algol
TRIANGULUM
Capella
AURIGA
ARIES
30°
60°
E

8L

July 6 at 1^h	July 21 at midnight
August 6 at 23^h	August 21 at 22^h
September 6 at 21^h	September 21 at 20^h
October 6 at 19^h	October 21 at 18^h
November 6 at 17^h	November 21 at 16^h

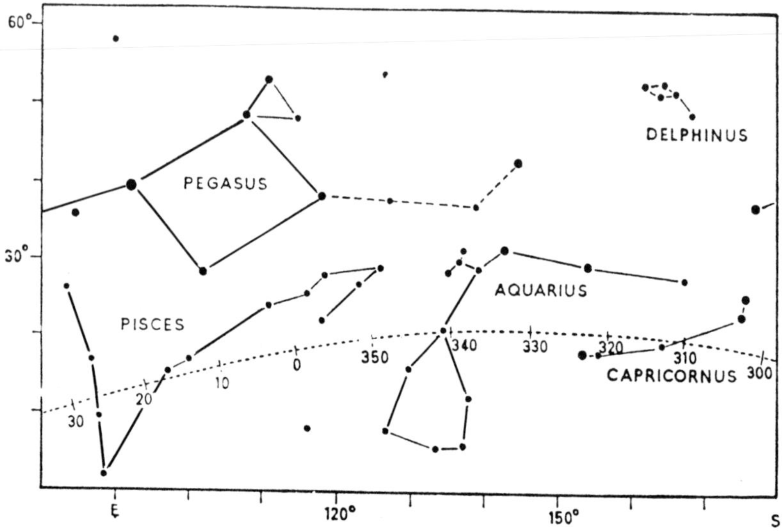

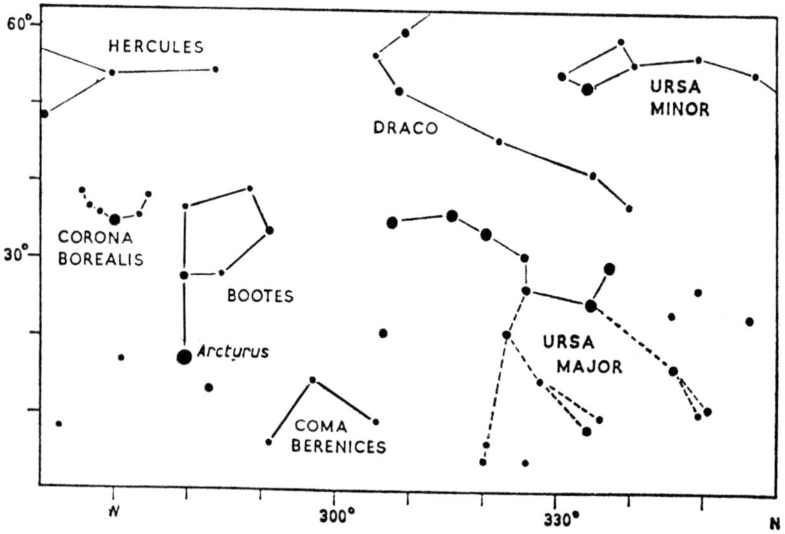

60°
30°
DELPHINUS
PEGASUS
AQUARIUS
PISCES
CAPRICORNUS
30 20 10 0 350 340 330 320 310 300
E 120° 150° S

60°
30°
HERCULES
DRACO
URSA MINOR
CORONA BOREALIS
BOOTES
Arcturus
URSA MAJOR
COMA BERENICES
W 300° 330° N

July 6 at 1^h	July 21 at midnight	**8R**
August 6 at 23^h	August 21 at 22^h	
September 6 at 21^h	September 21 at 20^h	
October 6 at 19^h	October 21 at 18^h	
November 6 at 17^h	November 21 at 16^h	

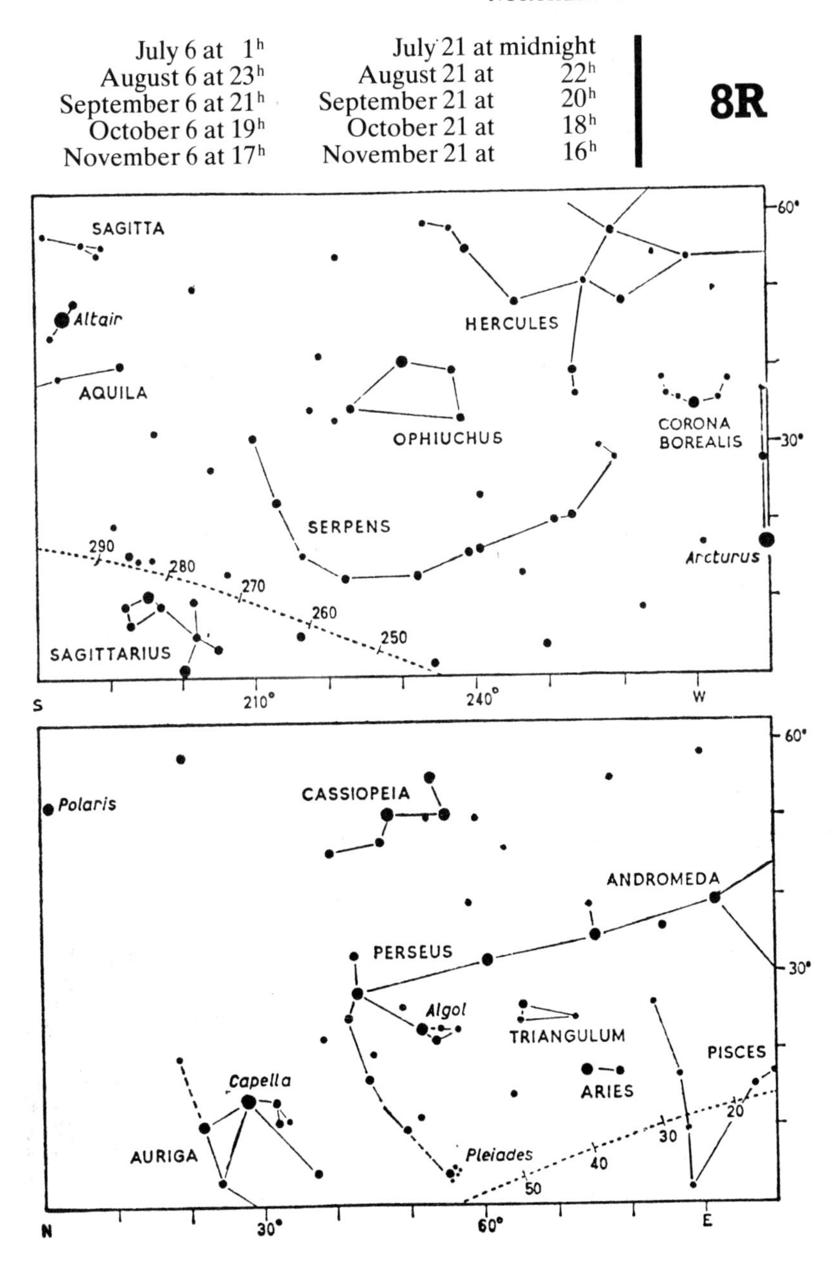

9L

August 6 at 1^h	August 21 at midnight
September 6 at 23^h	September 21 at 22^h
October 6 at 21^h	October 21 at 20^h
November 6 at 19^h	November 21 at 18^h
December 6 at 17^h	December 21 at 16^h

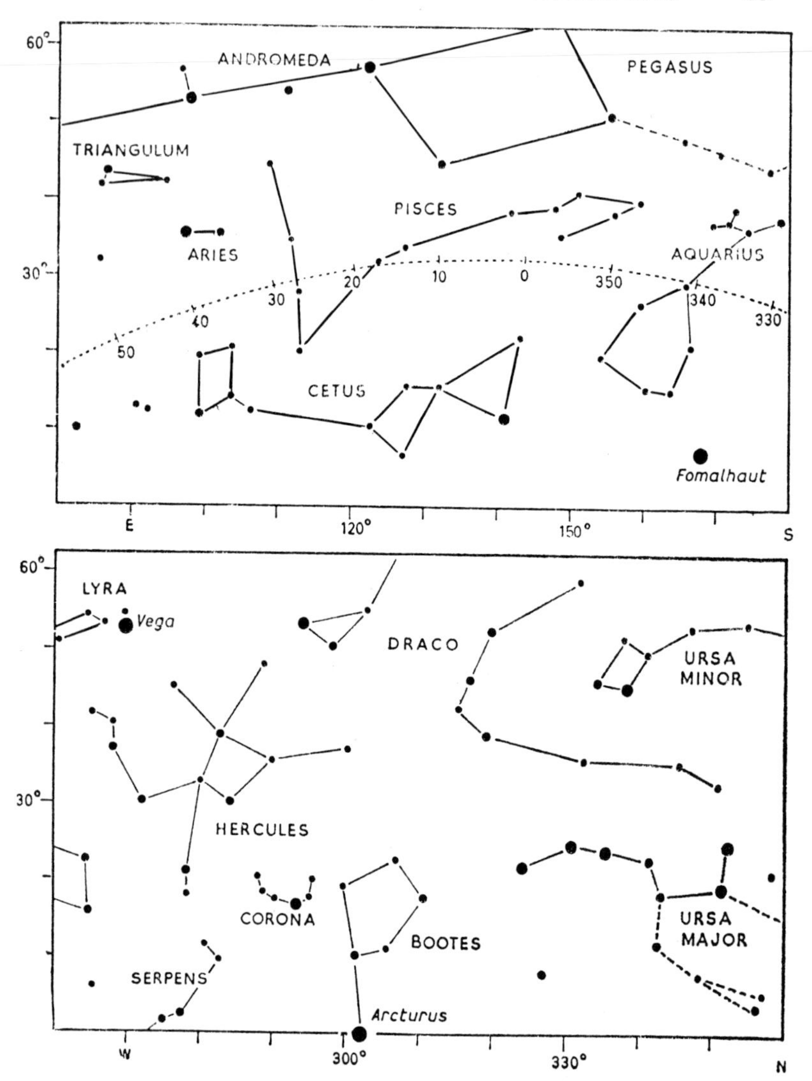

August 6 at 1h	August 21 at midnight	**9R**
September 6 at 23h	September 21 at 22h	
October 6 at 21h	October 21 at 20h	
November 6 at 19h	November 21 at 18h	
December 6 at 17h	December 21 at 16h	

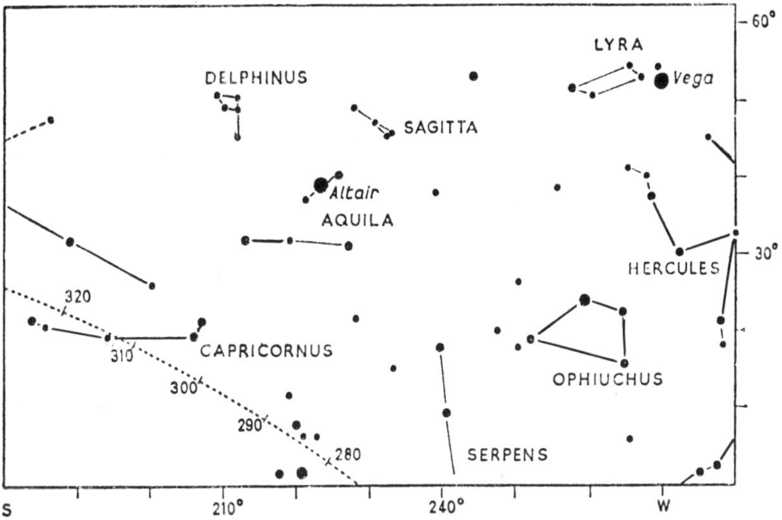

LYRA
Vega
DELPHINUS
SAGITTA
Altair
AQUILA
HERCULES
320
310
CAPRICORNUS
300
290
280
OPHIUCHUS
SERPENS
60°
30°
S
210°
240°
W

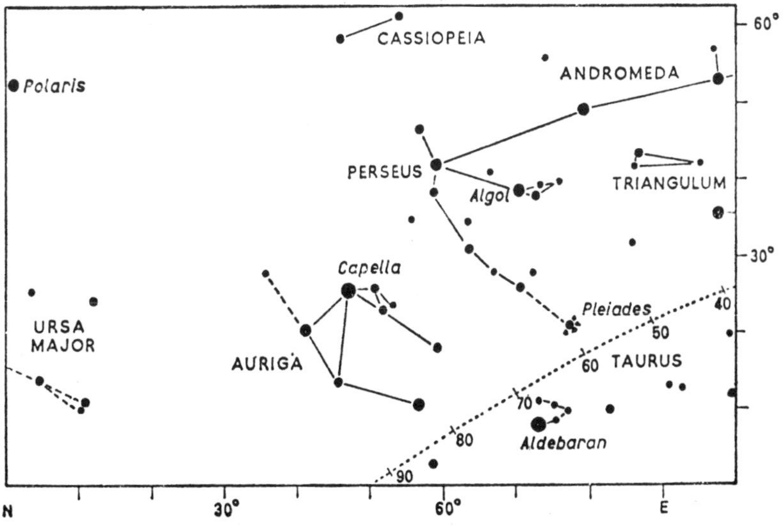

CASSIOPEIA
Polaris
ANDROMEDA
PERSEUS
Algol
TRIANGULUM
Capella
Pleiades
40
50
URSA MAJOR
AURIGA
60
TAURUS
70
80
Aldebaran
90
60°
30°
N
30°
60°
E

10L		
	August 6 at 3^h	August 21 at 2^h
	September 6 at 1^h	September 21 at midnight
	October 6 at 23^h	October 21 at 22^h
	November 6 at 21^h	November 21 at 20^h
	December 6 at 19^h	December 21 at 18^h

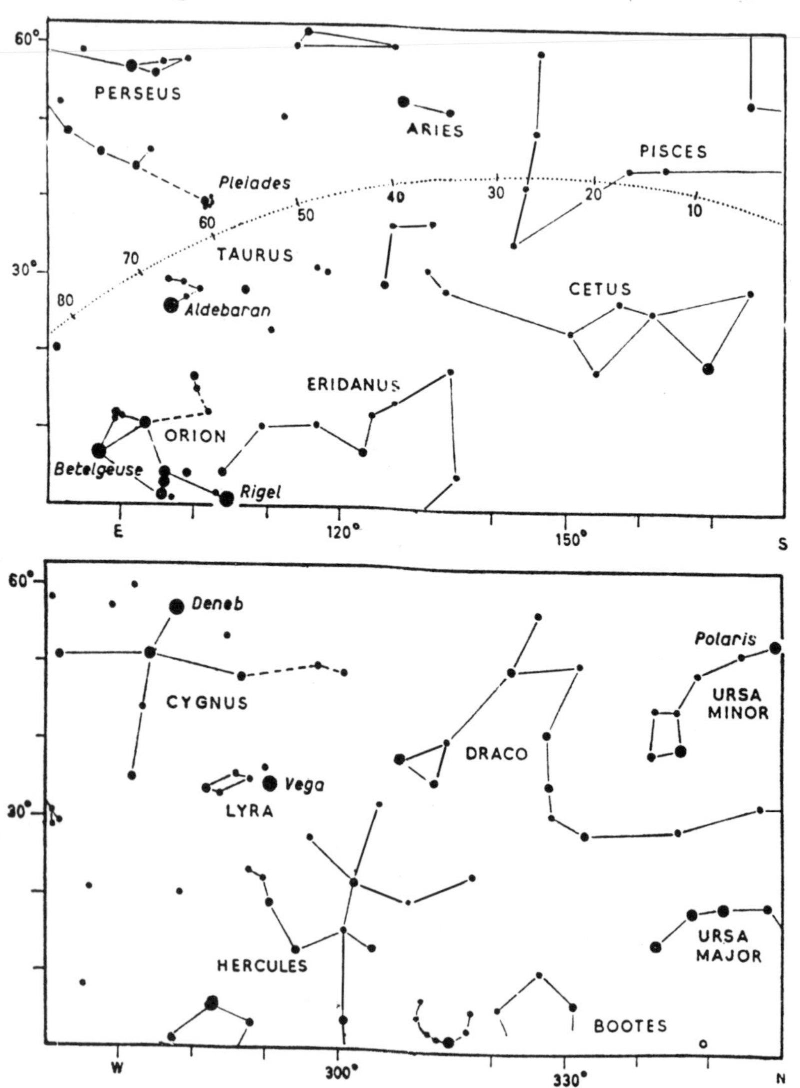

August 6 at 3h	August 21 at 2h	**10R**
September 6 at 1h	September 21 at midnight	
October 6 at 23h	October 21 at 22h	
November 6 at 21h	November 21 at 20h	
December 6 at 19h	December 21 at 18h	

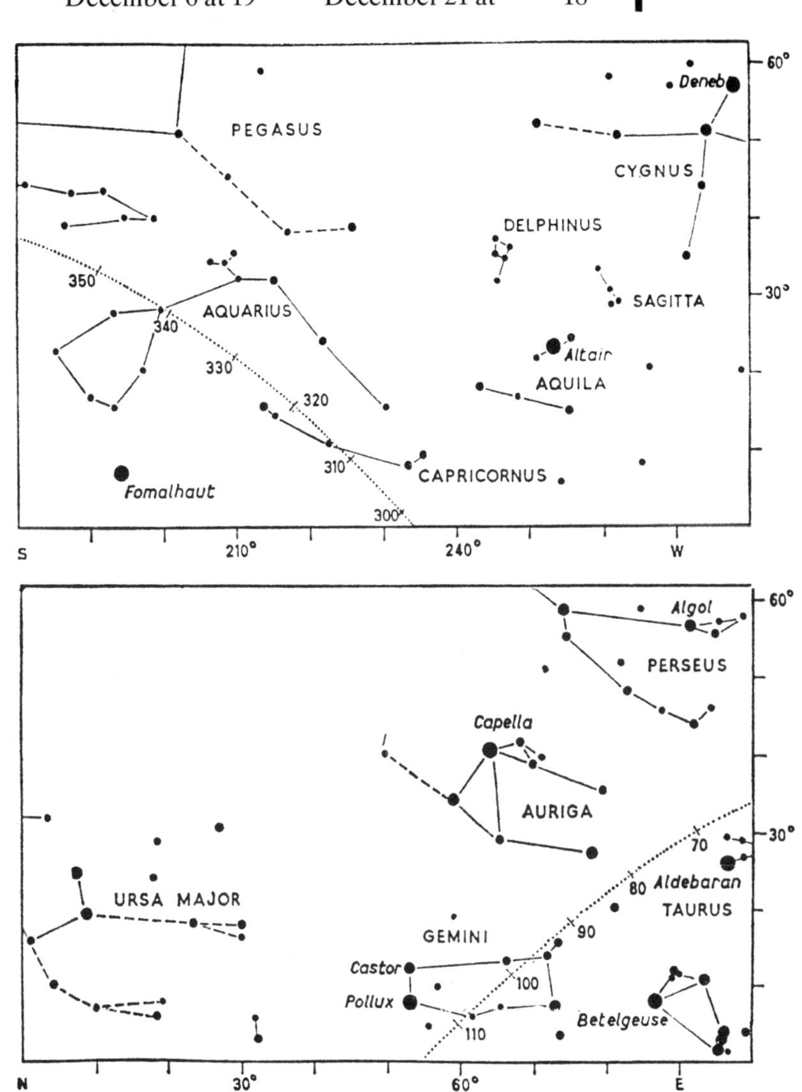

11L

September 6 at 3ʰ	September 21 at 2ʰ
October 6 at 1ʰ	October 21 at midnight
November 6 at 23ʰ	November 21 at 22ʰ
December 6 at 21ʰ	December 21 at 20ʰ
January 6 at 19ʰ	January 21 at 18ʰ

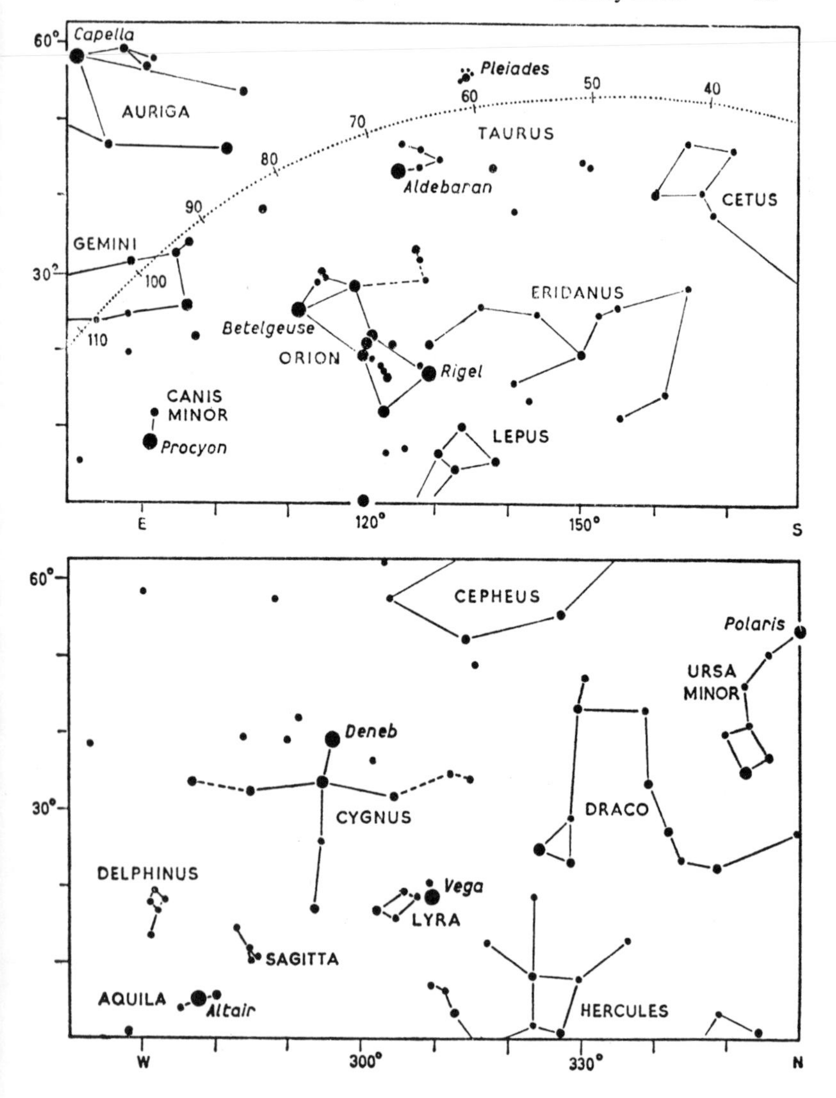

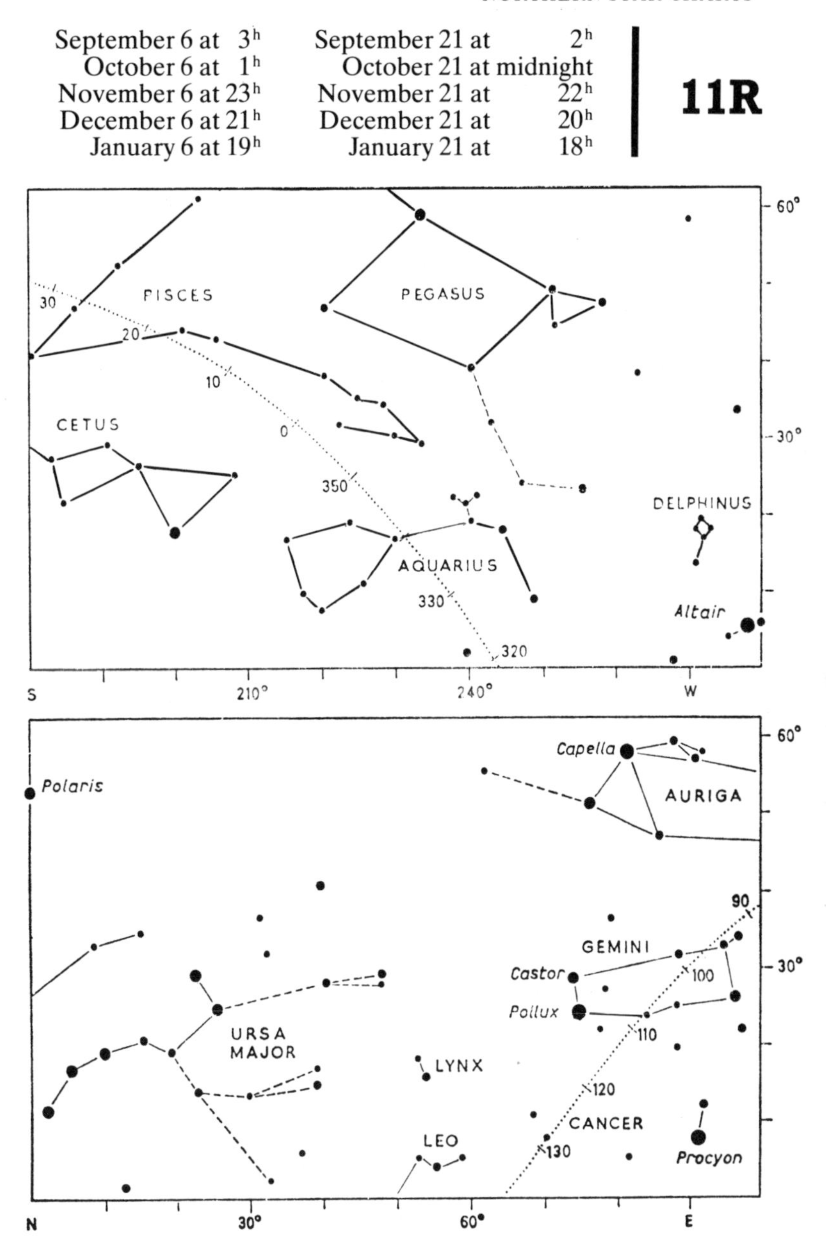
September 6 at 3h
October 6 at 1h
November 6 at 23h
December 6 at 21h
January 6 at 19h
September 21 at 2h
October 21 at midnight
November 21 at 22h
December 21 at 20h
January 21 at 18h
11R
PISCES
PEGASUS
CETUS
DELPHINUS
AQUARIUS
Altair
30
20
10
0
350
330
320
60°
30°
S
210°
240°
W
Capella
AURIGA
Polaris
GEMINI
Castor
Pollux
URSA
MAJOR
LYNX
CANCER
LEO
Procyon
90
100
110
120
130
N
30°
60°
E

12L		
	October 6 at 3h	October 21 at 2h
	November 6 at 1h	November 21 at midnight
	December 6 at 23h	December 21 at 22h
	January 6 at 21h	January 21 at 20h
	February 6 at 19h	February 21 at 18h

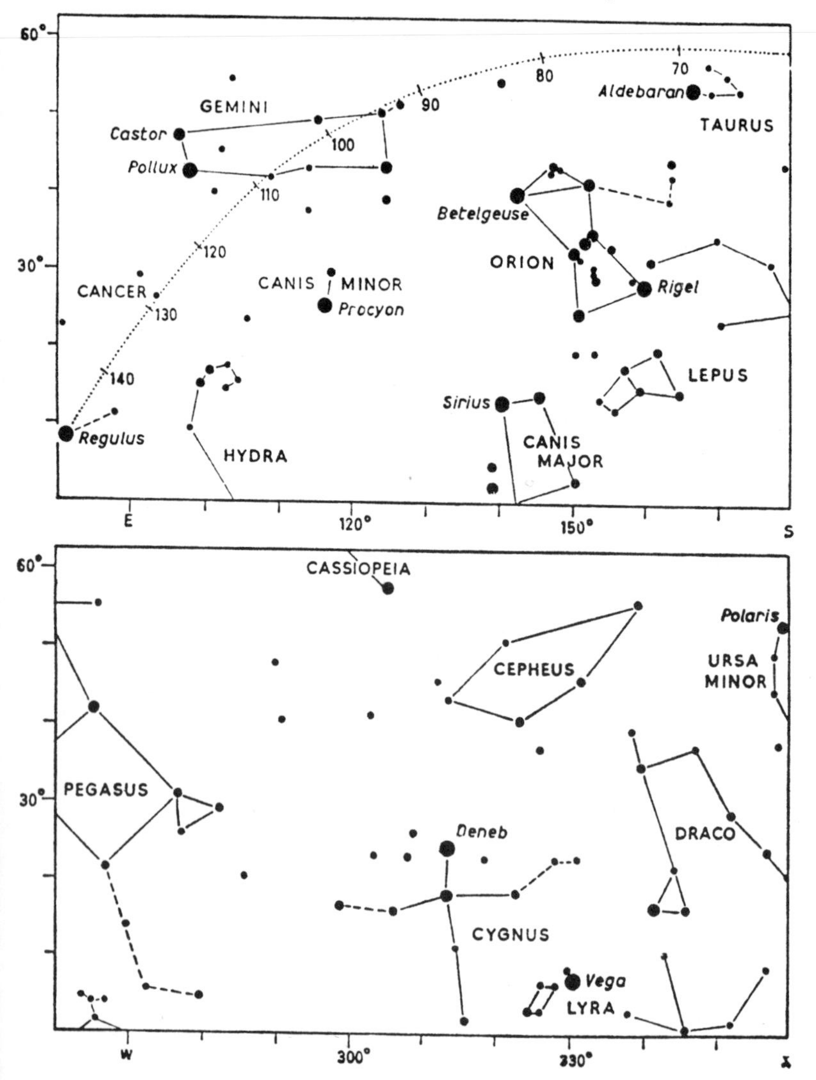

12R

October 6 at 3^h	October 21 at 2^h
November 6 at 1^h	November 21 at midnight
December 6 at 23^h	December 21 at 22^h
January 6 at 21^h	January 21 at 20^h
February 6 at 19^h	February 21 at 18^h

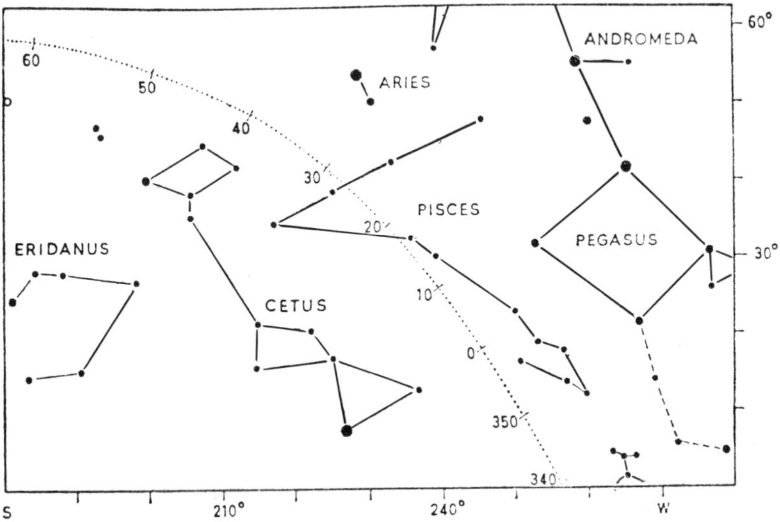

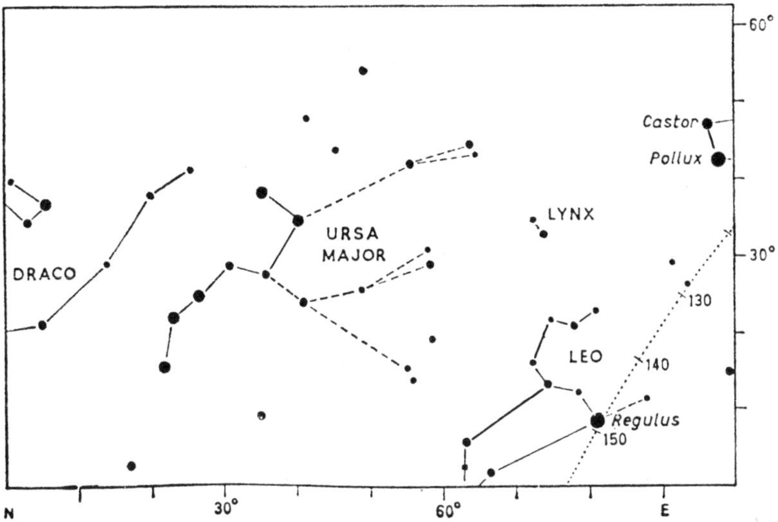

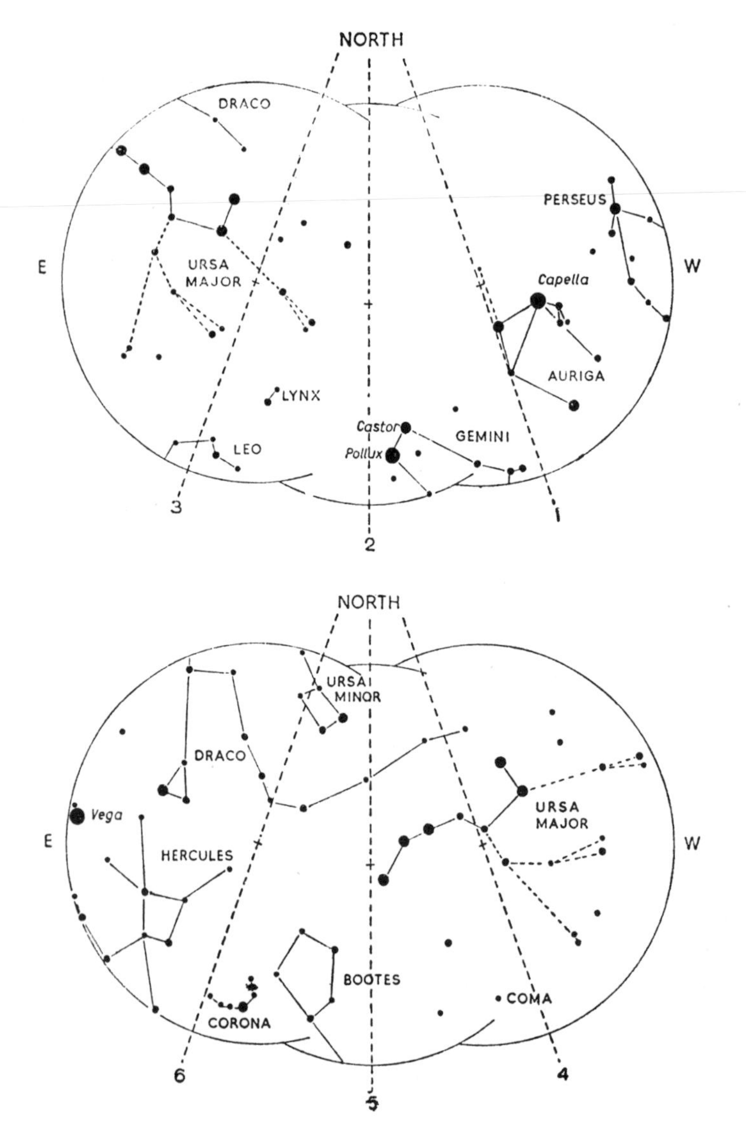

Northern Hemisphere Overhead Stars

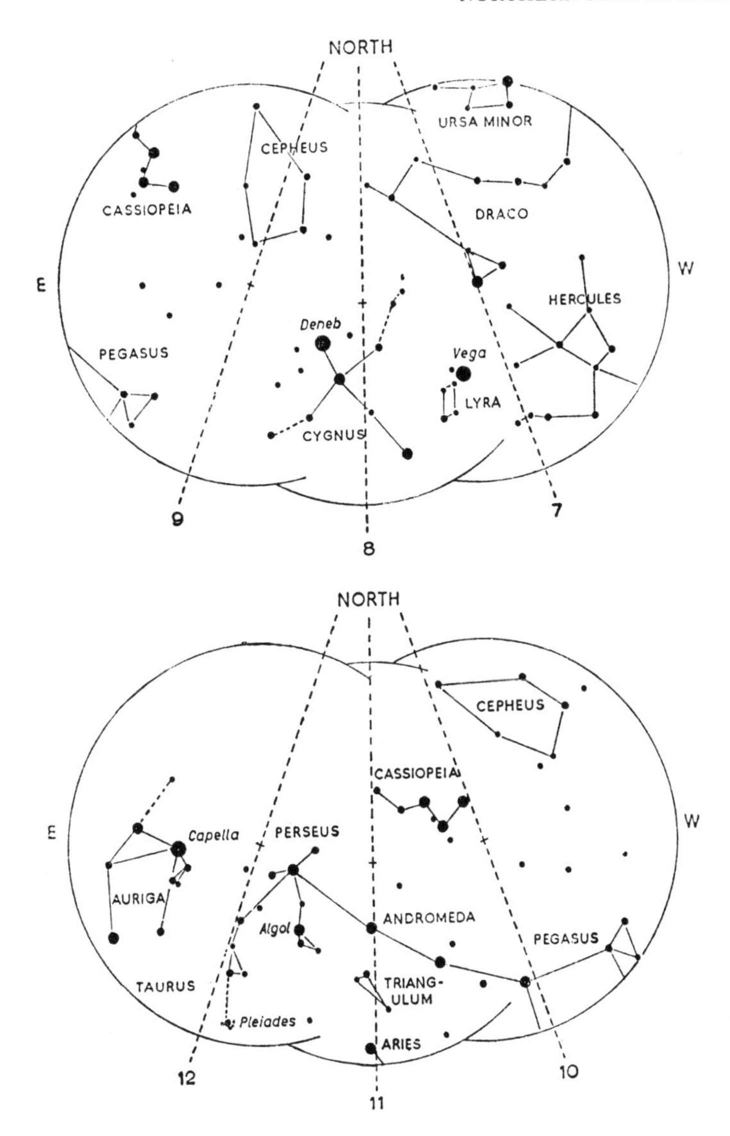

Northern Hemisphere Overhead Stars

1L

October 6 at 5h	October 21 at 4h
November 6 at 3h	November 21 at 2h
December 6 at 1h	December 21 at midnight
January 6 at 23h	January 21 at 22h
February 6 at 21h	February 21 at 20h

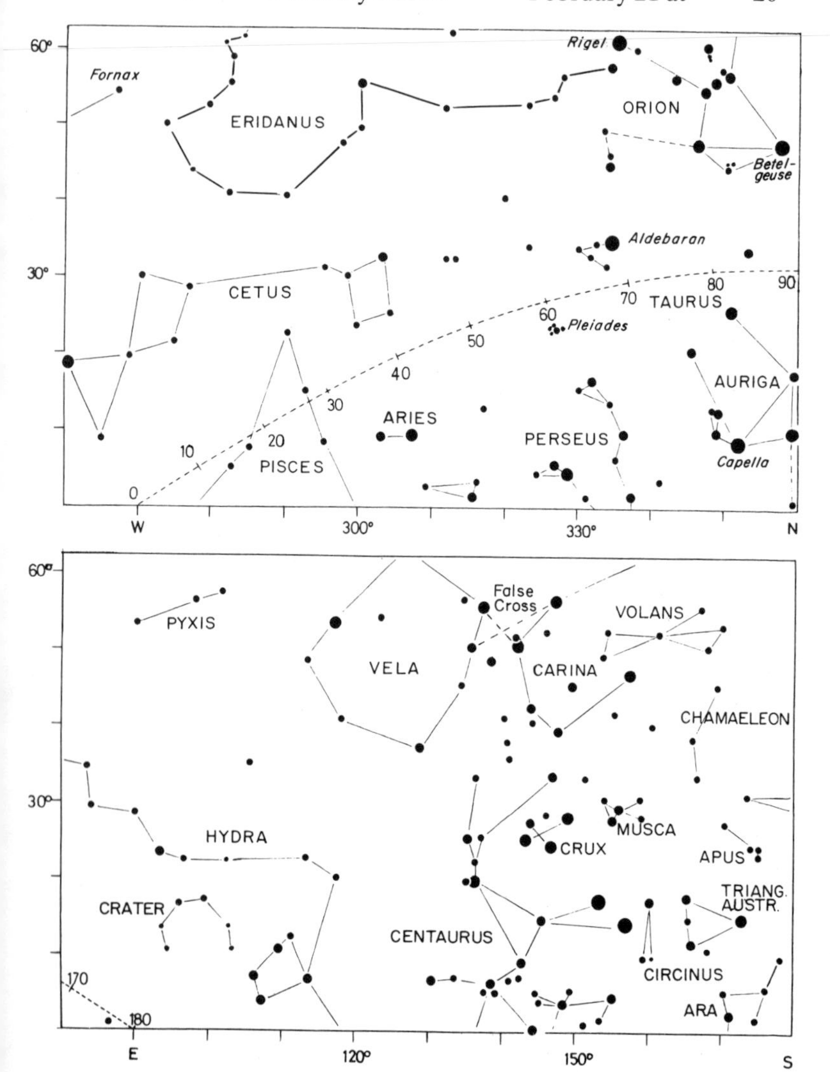

October 6 at 5^h / October 21 at 4^h
November 6 at 3^h / November 21 at 2^h
December 6 at 1^h / December 21 at midnight
January 6 at 23^h / January 21 at 22^h
February 6 at 21^h / February 21 at 20^h

1R

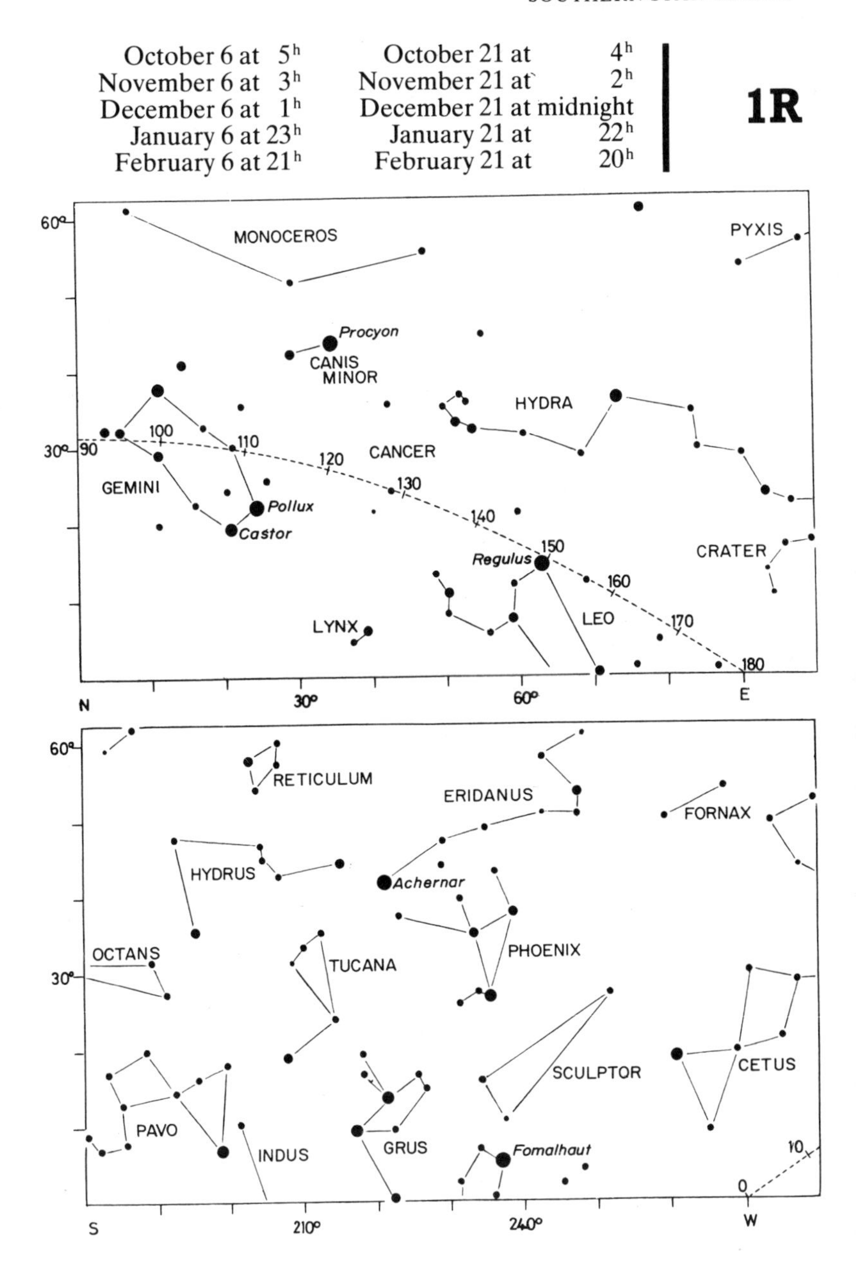

2L

November 6 at 5h	November 21 at 4h
December 6 at 3h	December 21 at 2h
January 6 at 1h	January 21 at midnight
February 6 at 23h	February 21 at 22h
March 6 at 21h	March 21 at 20h

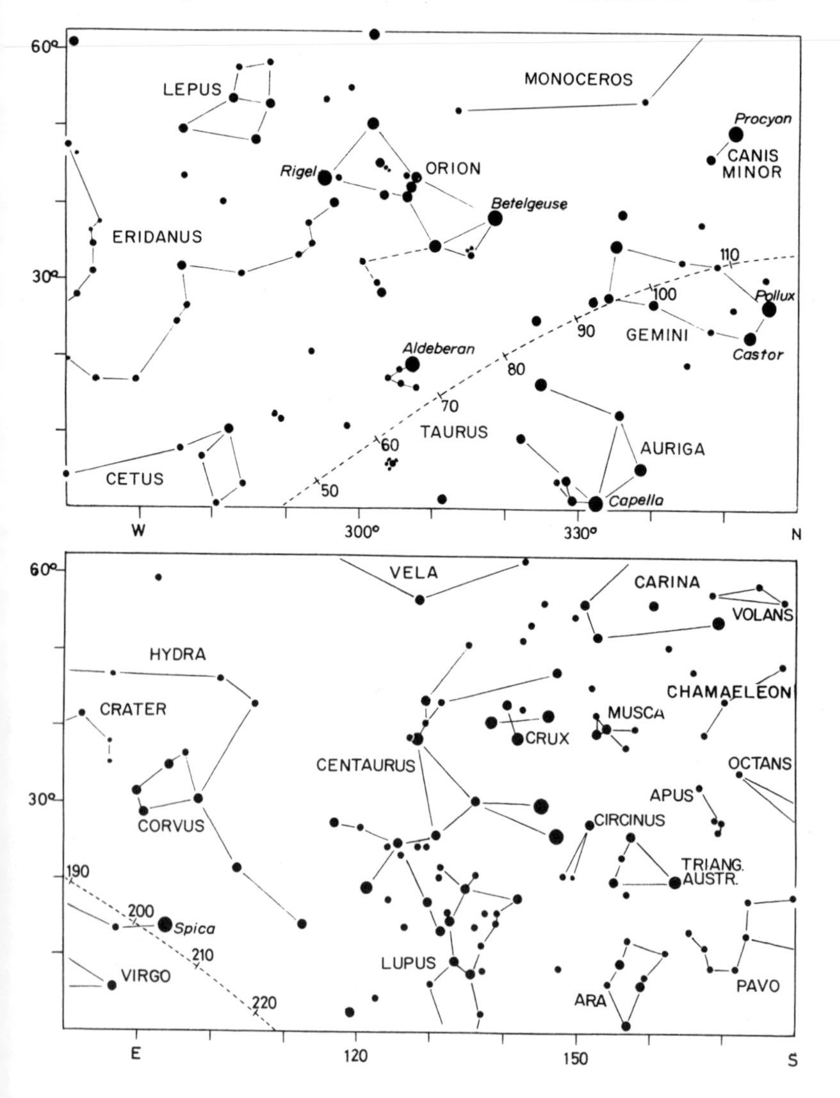

November 6 at 5ʰ	November 21 at 4ʰ
December 6 at 3ʰ	December 21 at 2ʰ
January 6 at 1ʰ	January 21 at midnight
February 6 at 23ʰ	February 21 at 22ʰ
March 6 at 21ʰ	March 21 at 20ʰ

2R

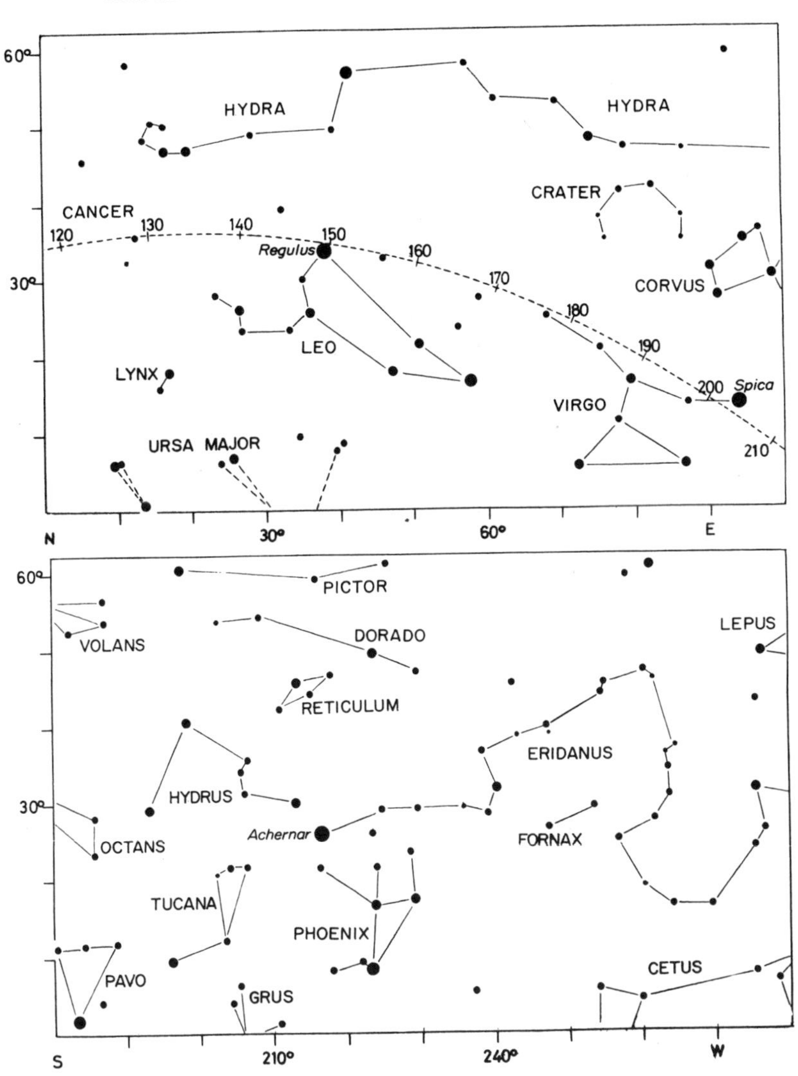

3L

January 6 at 3h	January 21 at 2h
February 6 at 1h	February 21 at midnight
March 6 at 23h	March 21 at 22h
April 6 at 21h	April 21 at 20h
May 6 at 19h	May 21 at 18h

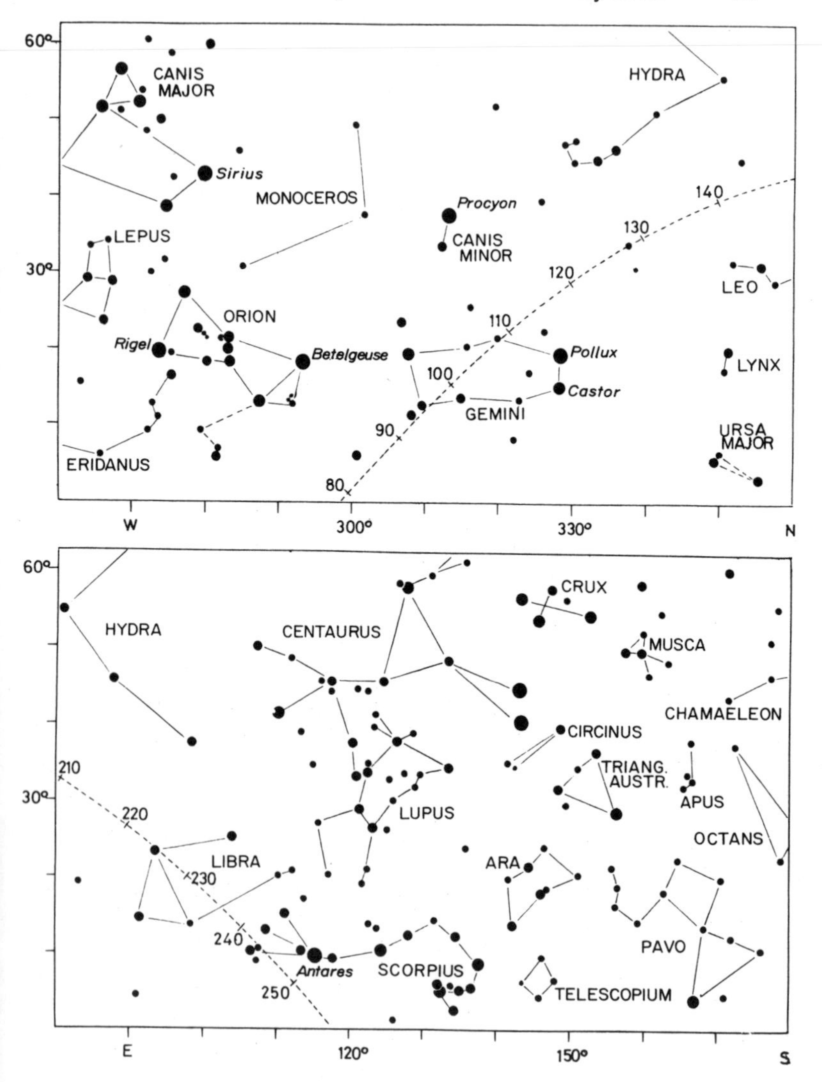

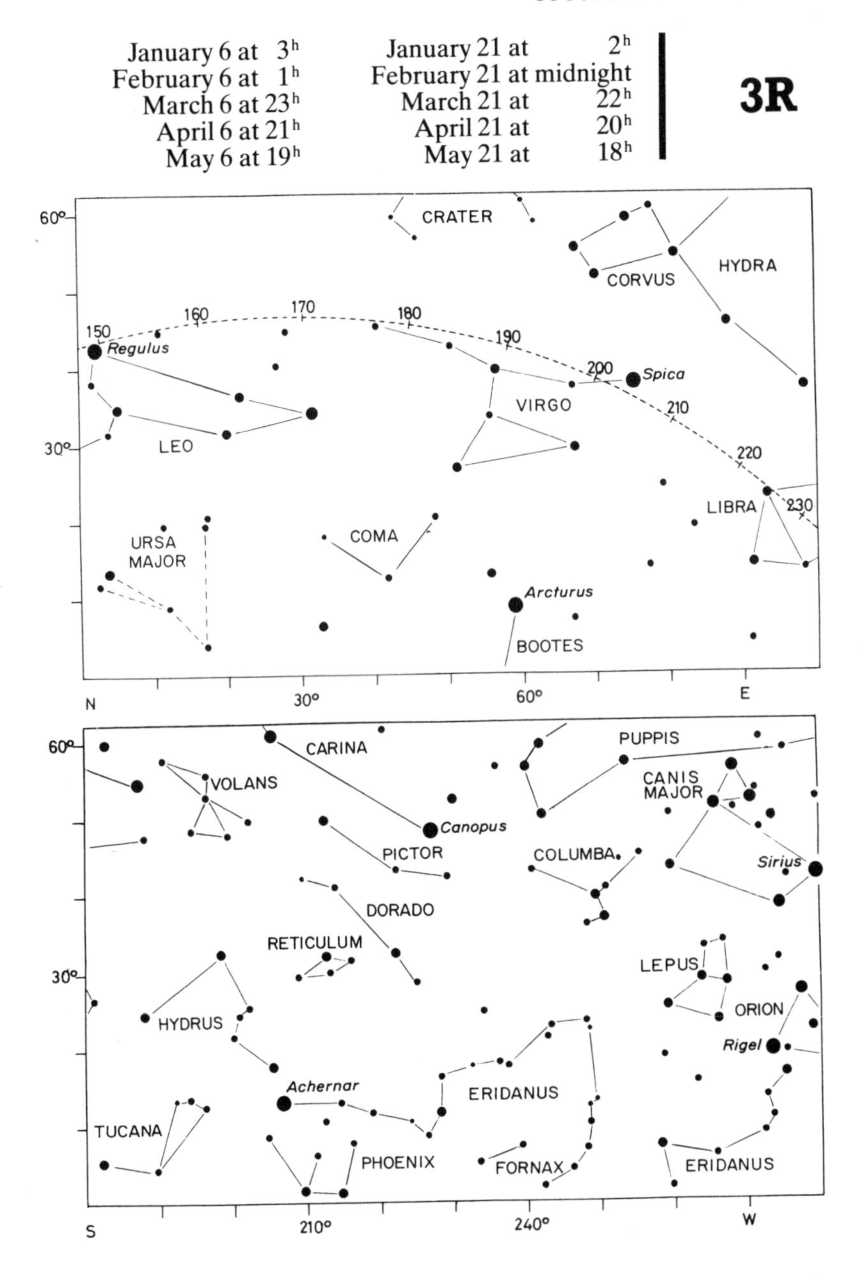
January 6 at 3h
February 6 at 1h
March 6 at 23h
April 6 at 21h
May 6 at 19h
January 21 at 2h
February 21 at midnight
March 21 at 22h
April 21 at 20h
May 21 at 18h
3R
CRATER
CORVUS
HYDRA
150
160
170
180
190
200
210
220
230
Regulus
Spica
VIRGO
LEO
LIBRA
URSA
MAJOR
COMA
Arcturus
BOOTES
60°
30°
N
30°
60°
E
CARINA
PUPPIS
VOLANS
CANIS
MAJOR
Canopus
PICTOR
COLUMBA
Sirius
DORADO
RETICULUM
LEPUS
ORION
HYDRUS
Rigel
Achernar
ERIDANUS
TUCANA
PHOENIX
FORNAX
ERIDANUS
S
210°
240°
W

4L

February 6 at 3^h	February 21 at 2^h
March 6 at 1^h	March 21 at midnight
April 6 at 23^h	April 21 at 22^h
May 6 at 21^h	May 21 at 20^h
June 6 at 19^h	June 21 at 18^h

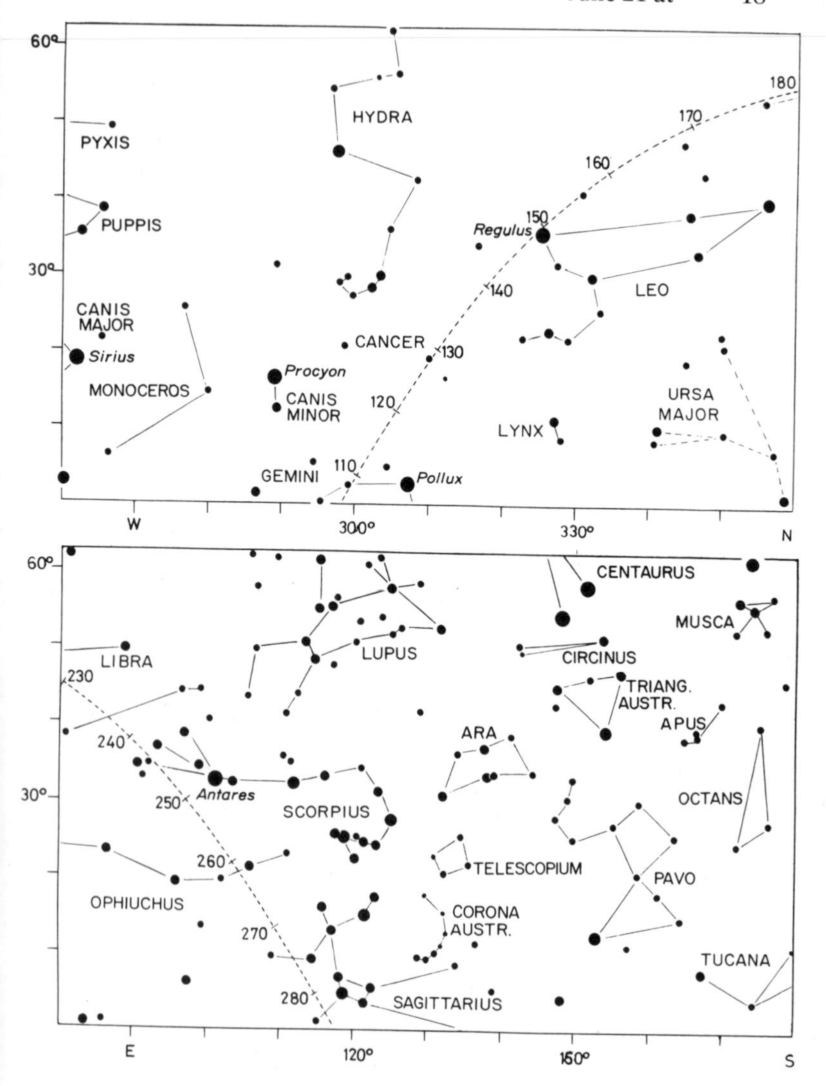

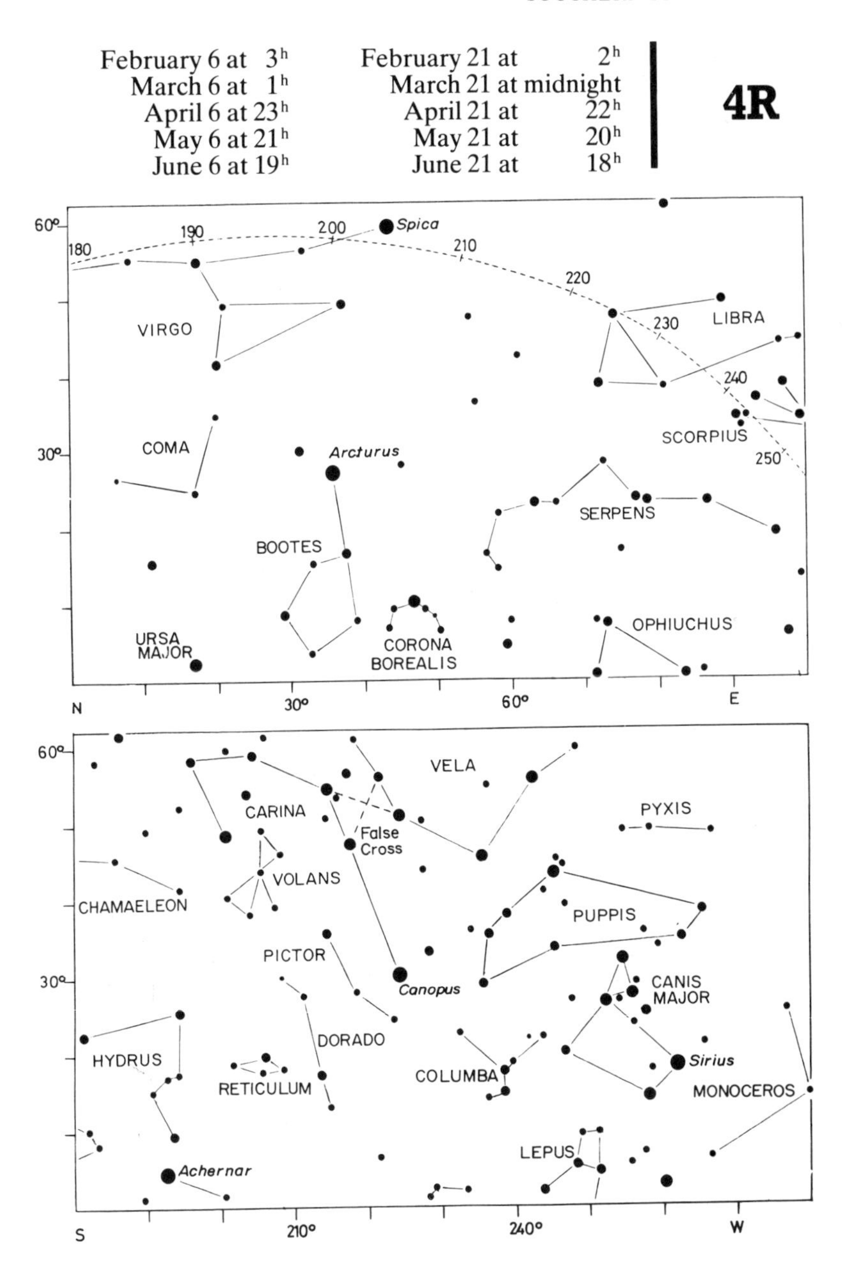
February 6 at 3h
March 6 at 1h
April 6 at 23h
May 6 at 21h
June 6 at 19h
February 21 at 2h
March 21 at midnight
April 21 at 22h
May 21 at 20h
June 21 at 18h
4R
60°
30°
180
190
200
210
220
230
240
250
Spica
VIRGO
LIBRA
SCORPIUS
COMA
Arcturus
SERPENS
BOOTES
OPHIUCHUS
URSA MAJOR
CORONA BOREALIS
N
30°
60°
E
VELA
CARINA
PYXIS
False Cross
VOLANS
CHAMAELEON
PUPPIS
PICTOR
Canopus
CANIS MAJOR
DORADO
HYDRUS
Sirius
RETICULUM
COLUMBA
MONOCEROS
LEPUS
Achernar
S
210°
240°
W

5L

March 6 at 3h	March 21 at 2h
April 6 at 1h	April 21 at midnight
May 6 at 23h	May 21 at 22h
June 6 at 21h	June 21 at 20h
July 6 at 19h	July 21 at 18h

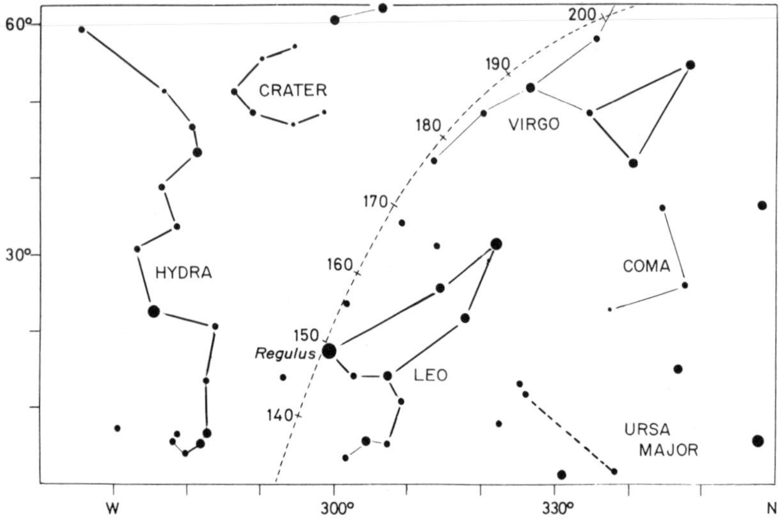

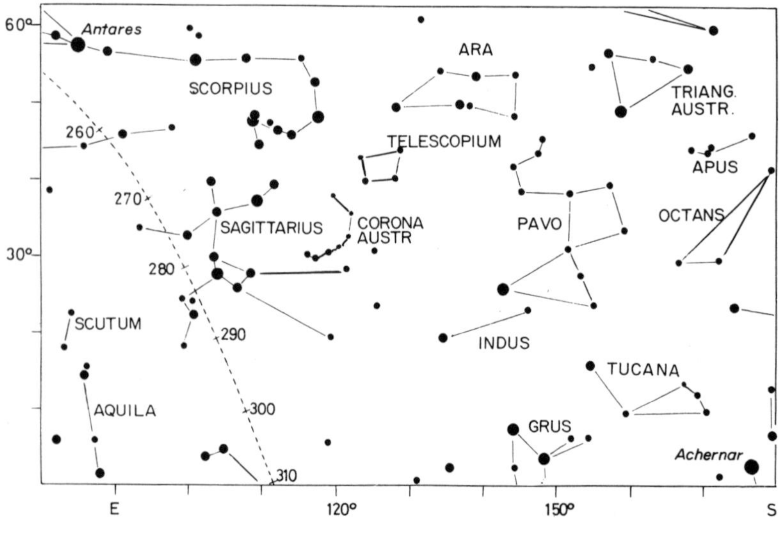

60°
CRATER
200
190
180
VIRGO
170
30°
HYDRA
160
COMA
150
Regulus
LEO
140
URSA MAJOR
W
300°
330°
N

60°
Antares
ARA
SCORPIUS
TRIANG. AUSTR.
260
TELESCOPIUM
APUS
270
OCTANS
SAGITTARIUS
CORONA AUSTR
PAVO
30°
280
SCUTUM
290
INDUS
TUCANA
AQUILA
300
GRUS
Achernar
310
E
120°
150°
S

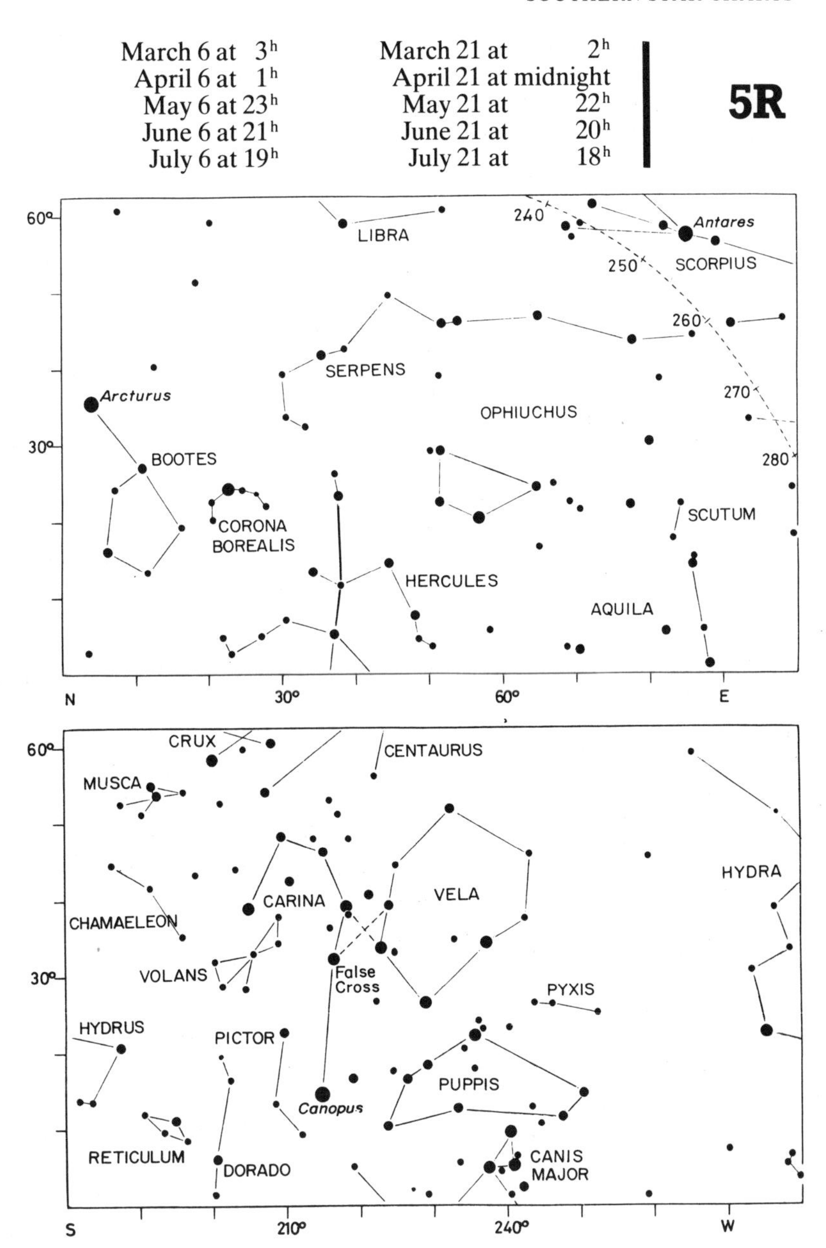
March 6 at 3h
April 6 at 1h
May 6 at 23h
June 6 at 21h
July 6 at 19h
March 21 at 2h
April 21 at midnight
May 21 at 22h
June 21 at 20h
July 21 at 18h
5R
60°
30°
N
30°
60°
E
LIBRA
240
Antares
250
SCORPIUS
260
SERPENS
270
Arcturus
OPHIUCHUS
280
BOOTES
CORONA
BOREALIS
SCUTUM
HERCULES
AQUILA
60°
30°
S
210°
240°
W
CRUX
CENTAURUS
MUSCA
HYDRA
VELA
CARINA
CHAMAELEON
VOLANS
False
Cross
PYXIS
HYDRUS
PICTOR
PUPPIS
Canopus
RETICULUM
DORADO
CANIS
MAJOR

6L

March 6 at 5ʰ	March 21 at 4ʰ
April 6 at 3ʰ	April 21 at 2ʰ
May 6 at 1ʰ	May 21 at midnight
June 6 at 23ʰ	June 21 at 22ʰ
July 6 at 21ʰ	July 21 at 20ʰ

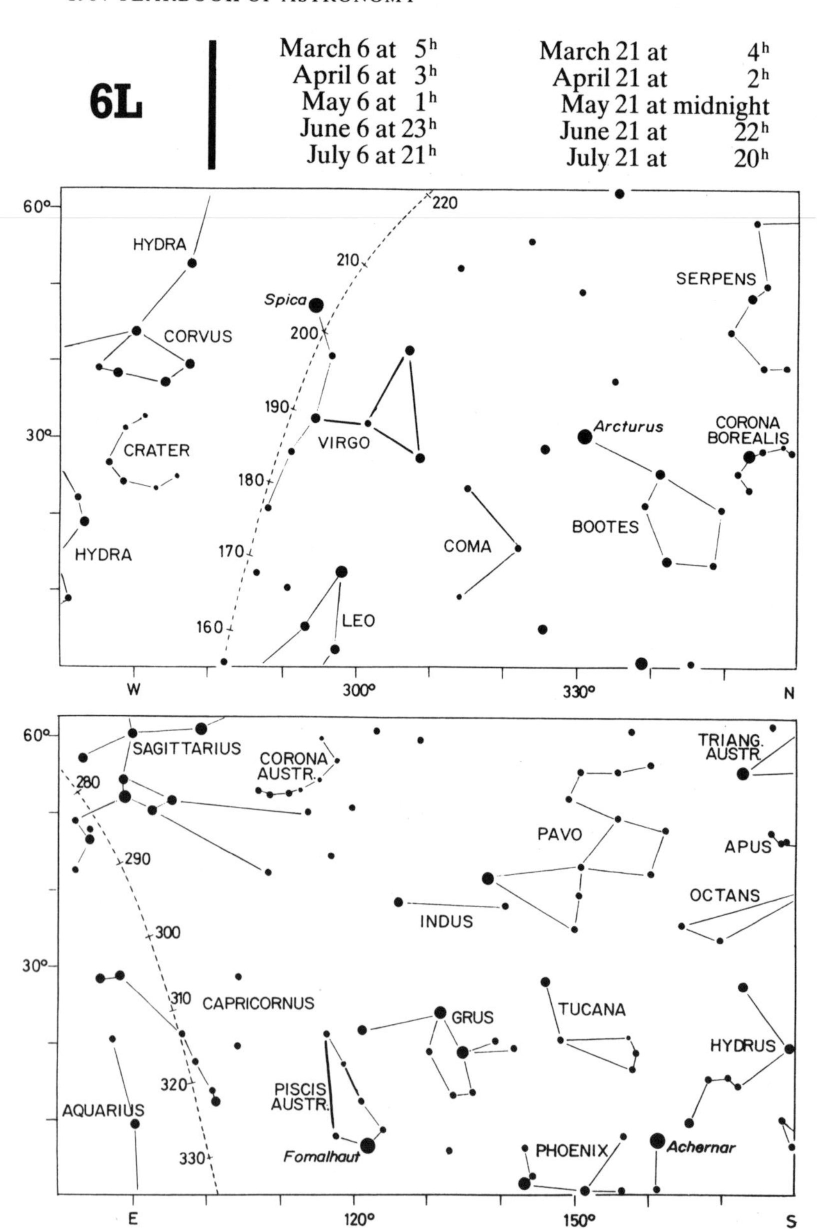

March 6 at 5h	March 21 at 4h	**6R**
April 6 at 3h	April 21 at 2h	
May 6 at 1h	May 21 at midnight	
June 6 at 23h	June 21 at 22h	
July 6 at 21h	July 21 at 20h	

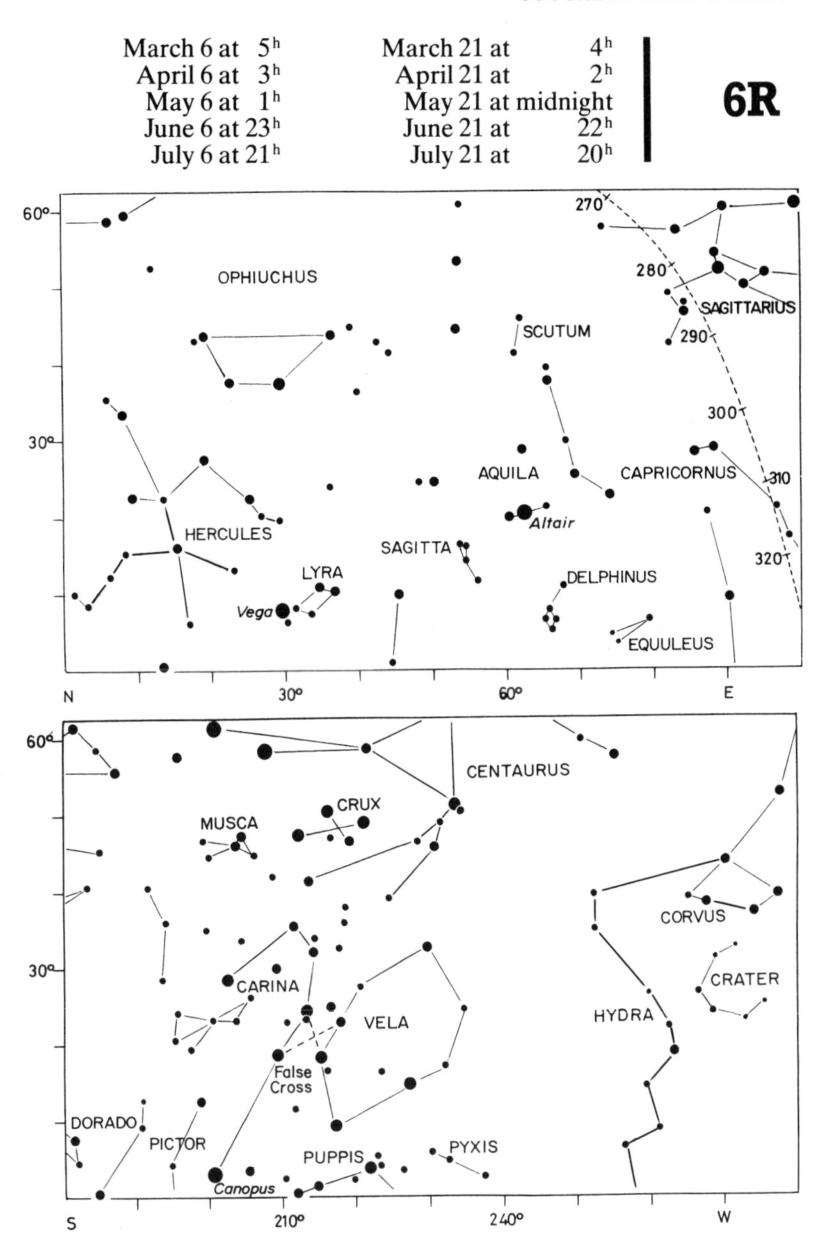

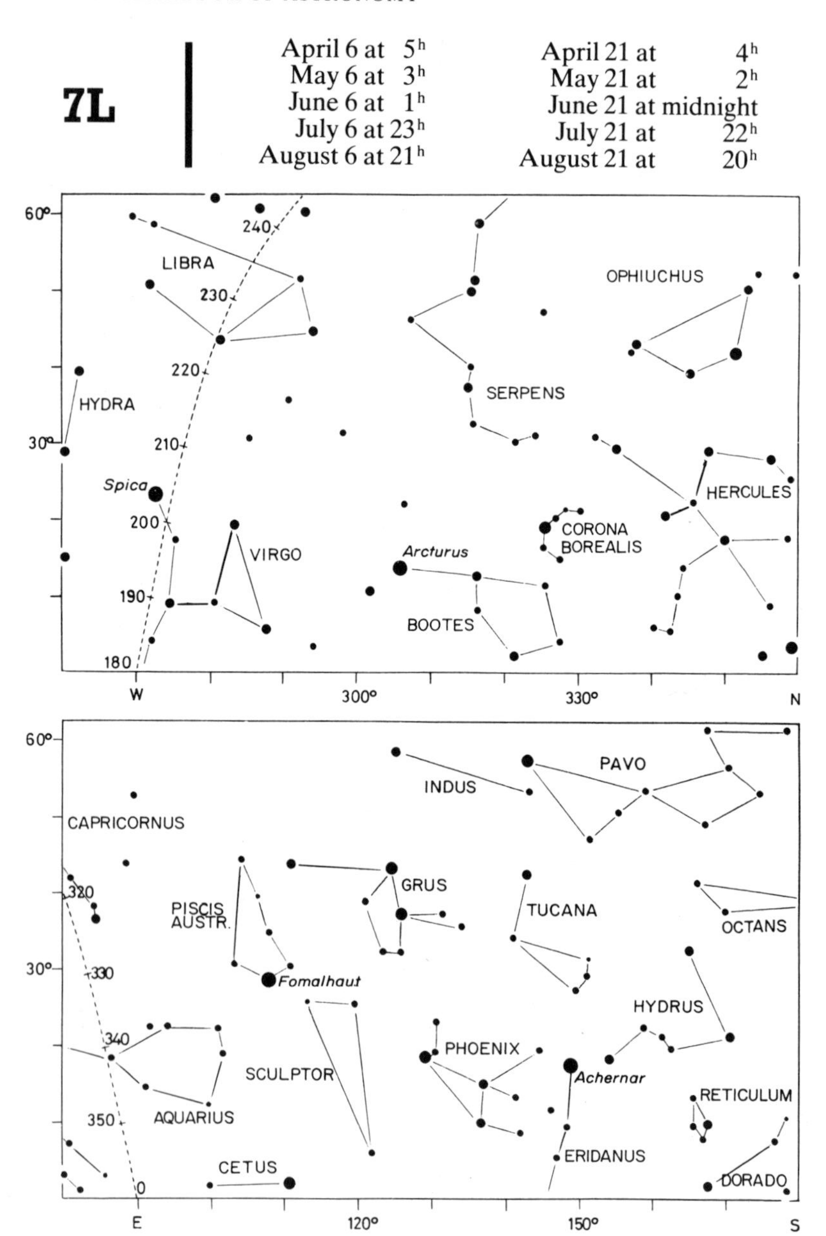
7L
April 6 at 5h
May 6 at 3h
June 6 at 1h
July 6 at 23h
August 6 at 21h
April 21 at 4h
May 21 at 2h
June 21 at midnight
July 21 at 22h
August 21 at 20h
60°
30°
240
230
220
210
200
190
180
LIBRA
HYDRA
Spica
VIRGO
Arcturus
BOOTES
SERPENS
OPHIUCHUS
CORONA BOREALIS
HERCULES
W
300°
330°
N
60°
30°
CAPRICORNUS
320
330
340
350
0
PISCIS AUSTR.
Fomalhaut
AQUARIUS
CETUS
SCULPTOR
INDUS
GRUS
PHOENIX
PAVO
TUCANA
Achernar
ERIDANUS
HYDRUS
OCTANS
RETICULUM
DORADO
E
120°
150°
S

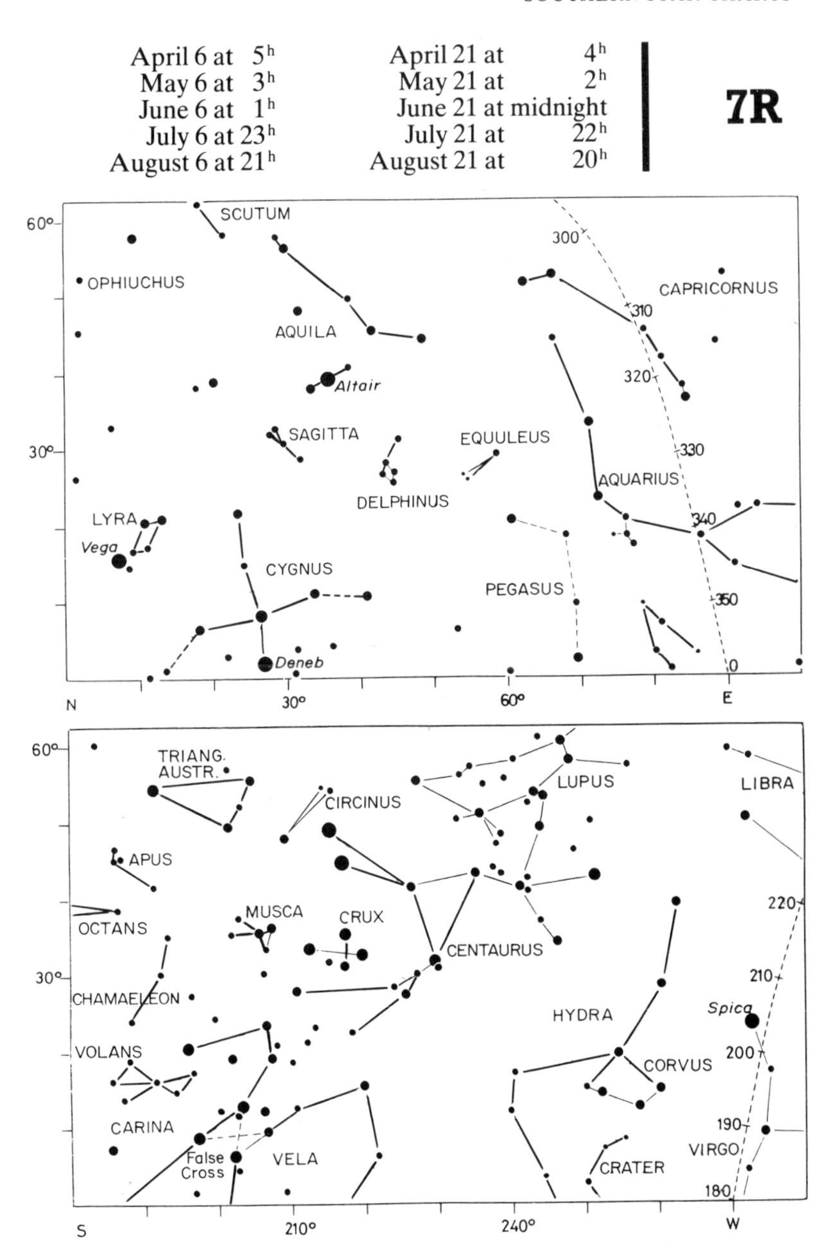
April 6 at 5h
May 6 at 3h
June 6 at 1h
July 6 at 23h
August 6 at 21h
April 21 at 4h
May 21 at 2h
June 21 at midnight
July 21 at 22h
August 21 at 20h
7R
60°
30°
SCUTUM
OPHIUCHUS
AQUILA
Altair
SAGITTA
DELPHINUS
EQUULEUS
LYRA
Vega
CYGNUS
Deneb
PEGASUS
AQUARIUS
CAPRICORNUS
300
310
320
330
340
350
0
N
30°
60°
E
TRIANG.
AUSTR.
APUS
OCTANS
CHAMAELEON
VOLANS
CARINA
False
Cross
VELA
MUSCA
CRUX
CIRCINUS
CENTAURUS
LUPUS
LIBRA
HYDRA
CORVUS
CRATER
VIRGO
Spica
220
210
200
190
180
S
210°
240°
W

8L

May 6 at 5h	May 21 at 4h
June 6 at 3h	June 21 at 2h
July 6 at 1h	July 21 at midnight
August 6 at 23h	August 21 at 22h
September 6 at 21h	September 21 at 20h

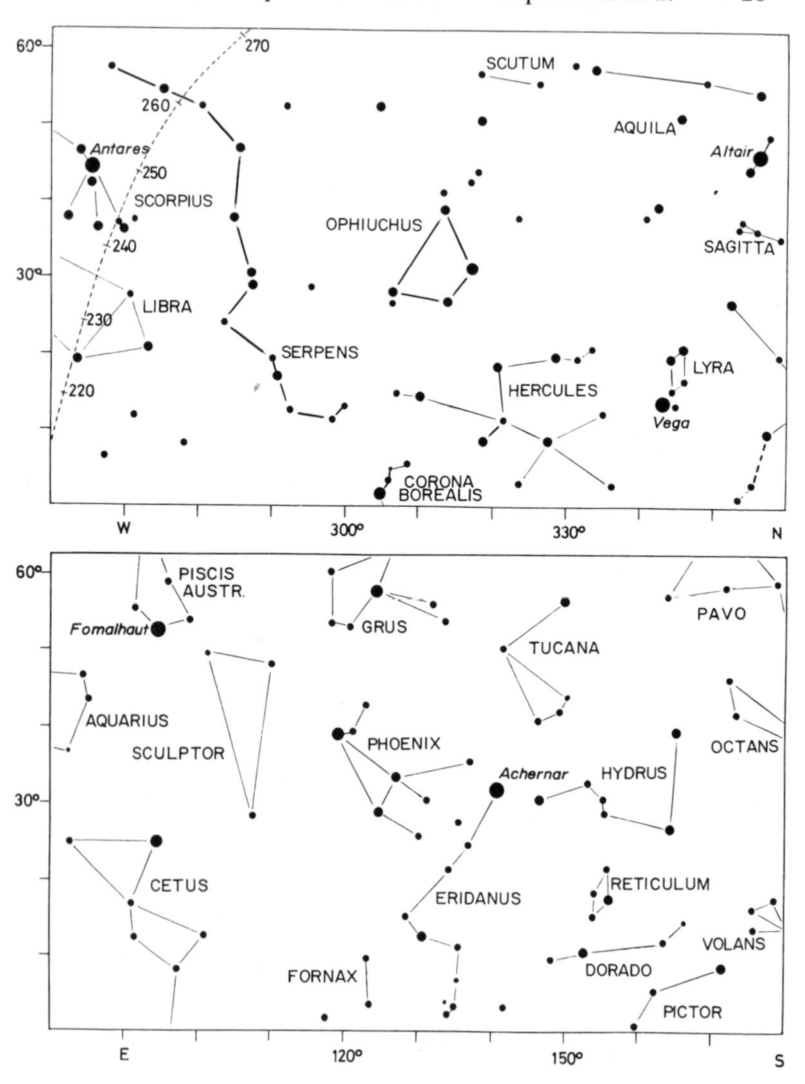

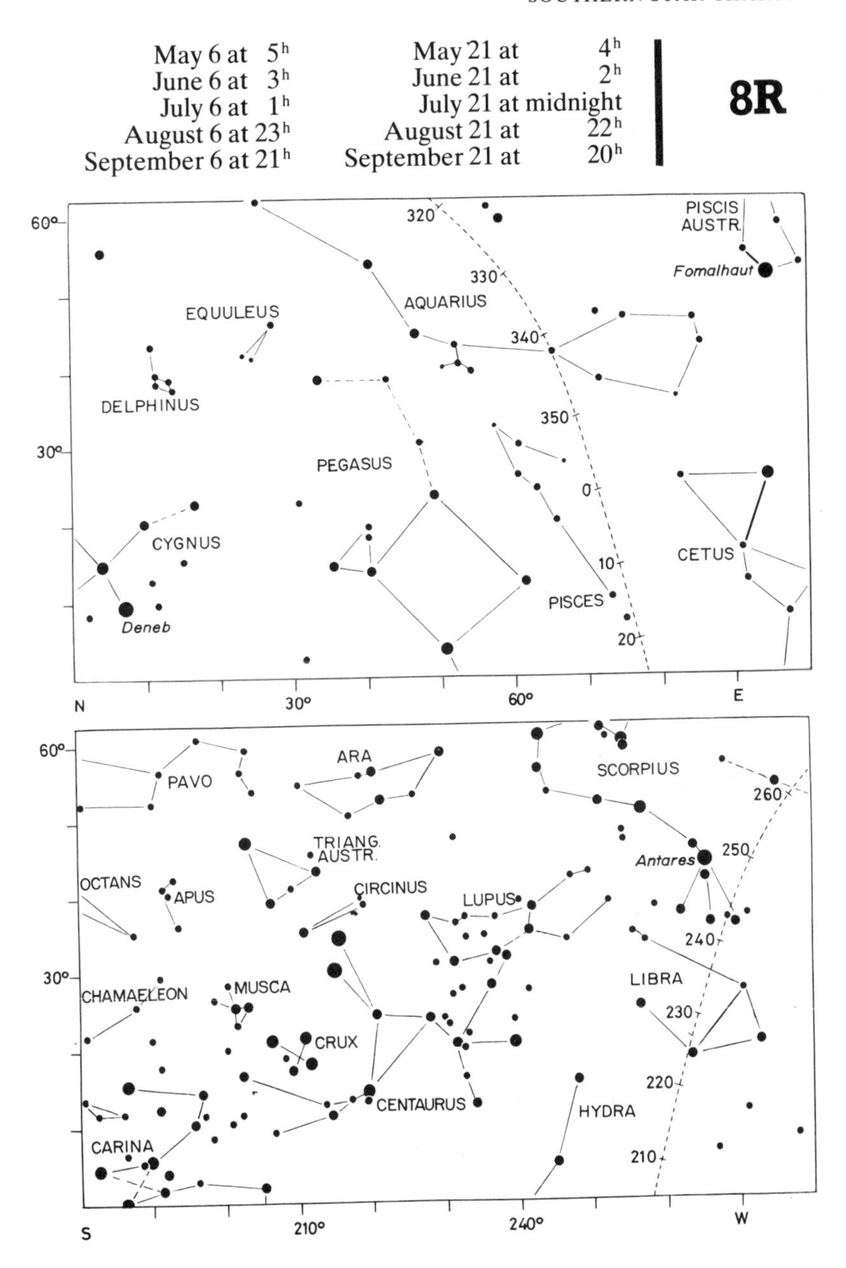
May 6 at 5h
June 6 at 3h
July 6 at 1h
August 6 at 23h
September 6 at 21h
May 21 at 4h
June 21 at 2h
July 21 at midnight
August 21 at 22h
September 21 at 20h
8R
PISCIS AUSTR.
Fomalhaut
EQUULEUS
AQUARIUS
DELPHINUS
PEGASUS
CYGNUS
Deneb
CETUS
PISCES
320
330
340
350
0
10
20
60°
30°
N
30°
60°
E
ARA
PAVO
SCORPIUS
TRIANG. AUSTR.
Antares
OCTANS
APUS
CIRCINUS
LUPUS
CHAMAELEON
MUSCA
LIBRA
CRUX
CENTAURUS
HYDRA
CARINA
260
250
240
230
220
210
S
210°
240°
W

9L

June 6 at 5^h	June 21 at 4^h
July 6 at 3^h	July 21 at 2^h
August 6 at 1^h	August 21 at midnight
September 6 at 23^h	September 21 at 22^h
October 6 at 21^h	October 21 at 20^h

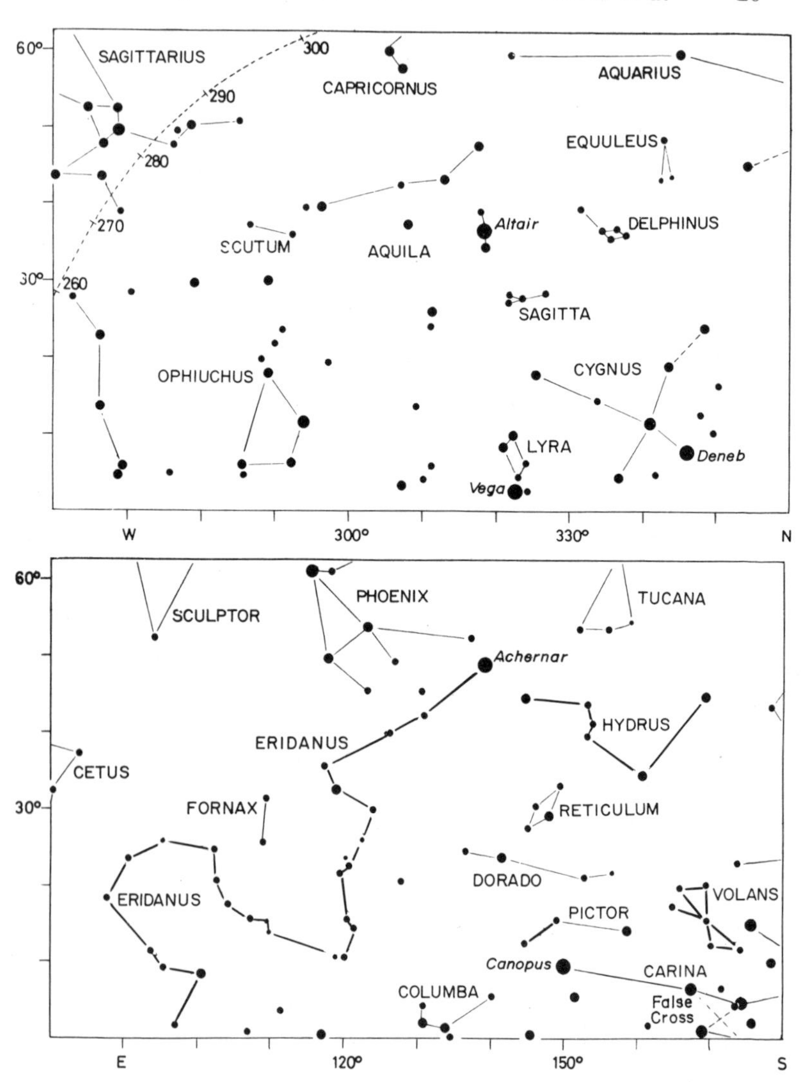

June 6 at 5^h	June 21 at 4^h
July 6 at 3^h	July 21 at 2^h
August 6 at 1^h	August 21 at midnight
September 6 at 23^h	September 21 at 22^h
October 6 at 21^h	October 21 at 20^h

9R

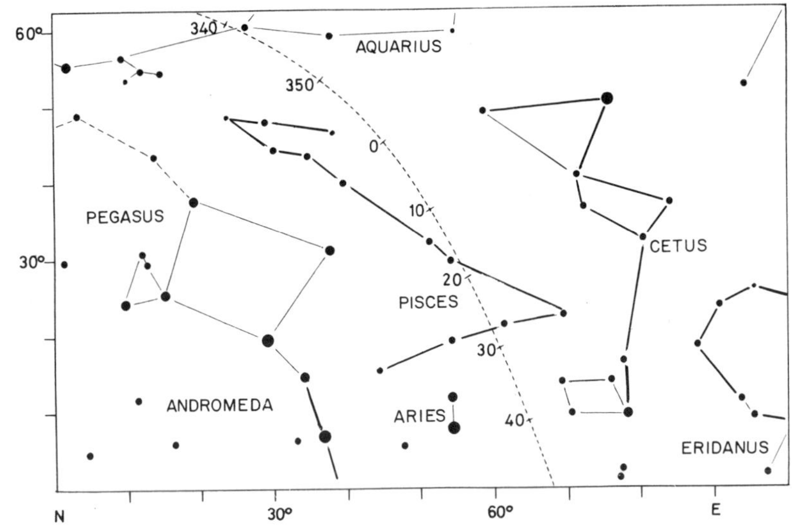

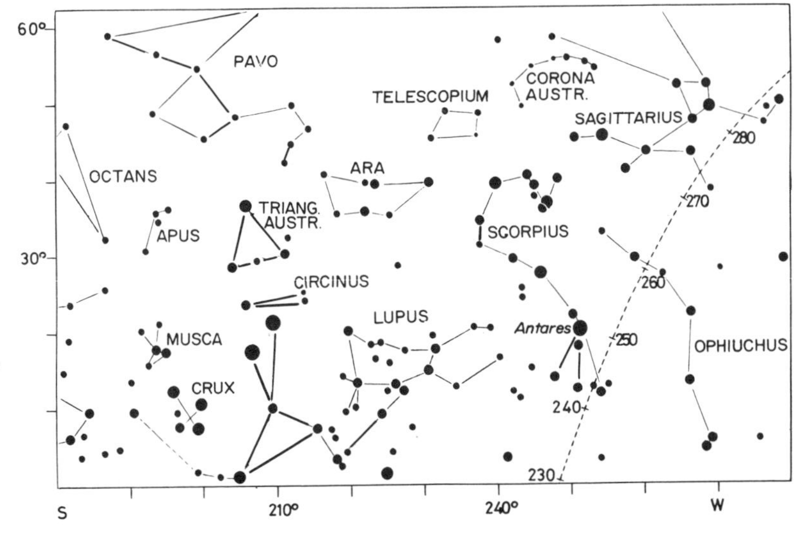

60° 340 AQUARIUS 350 0 10 PEGASUS 30° 20 CETUS PISCES 30 ANDROMEDA ARIES 40 ERIDANUS
N 30° 60° E

60° PAVO CORONA AUSTR. TELESCOPIUM SAGITTARIUS 280 ARA OCTANS TRIANG. AUSTR. 270 APUS SCORPIUS 30° CIRCINUS 260 LUPUS Antares 250 MUSCA OPHIUCHUS CRUX 240 230
S 210° 240° W

10L

July 6 at 5h	July 21 at 4h
August 6 at 3h	August 21 at 2h
September 6 at 1h	September 21 at midnight
October 6 at 23h	October 21 at 22h
November 6 at 21h	November 21 at 20h

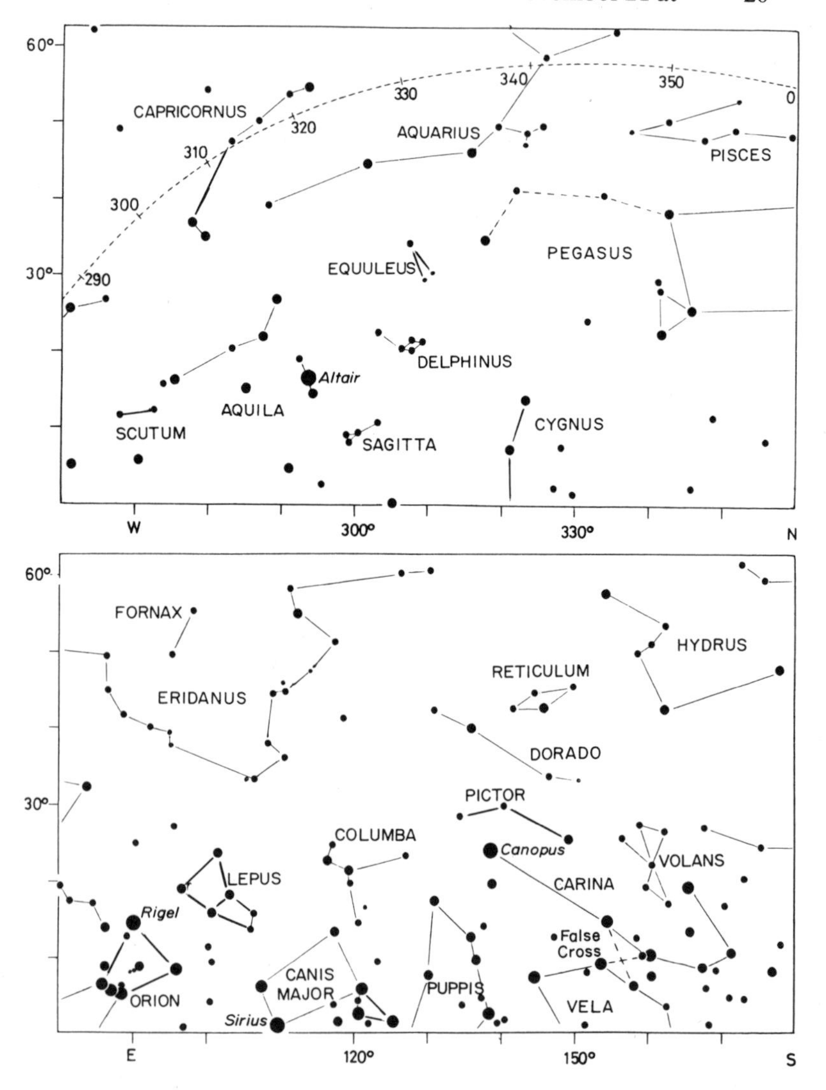

July 6 at 5^h	July 21 at 4^h		**10R**
August 6 at 3^h	August 21 at 2^h		
September 6 at 1^h	September 21 at midnight		
October 6 at 23^h	October 21 at 22^h		
November 6 at 21^h	November 21 at 20^h		

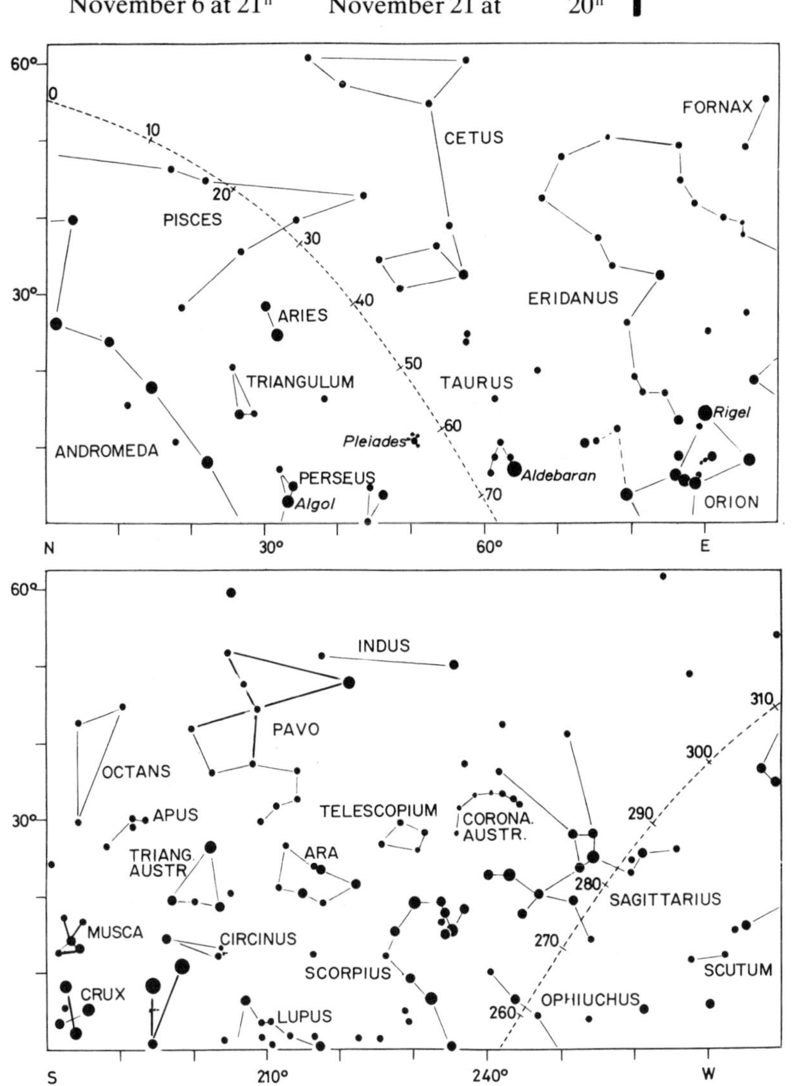

11L

August 6 at 5h	August 21 at 4h
September 6 at 3h	September 21 at 2h
October 6 at 1h	October 21 at midnight
November 6 at 23h	November 21 at 22h
December 6 at 21h	December 21 at 20h

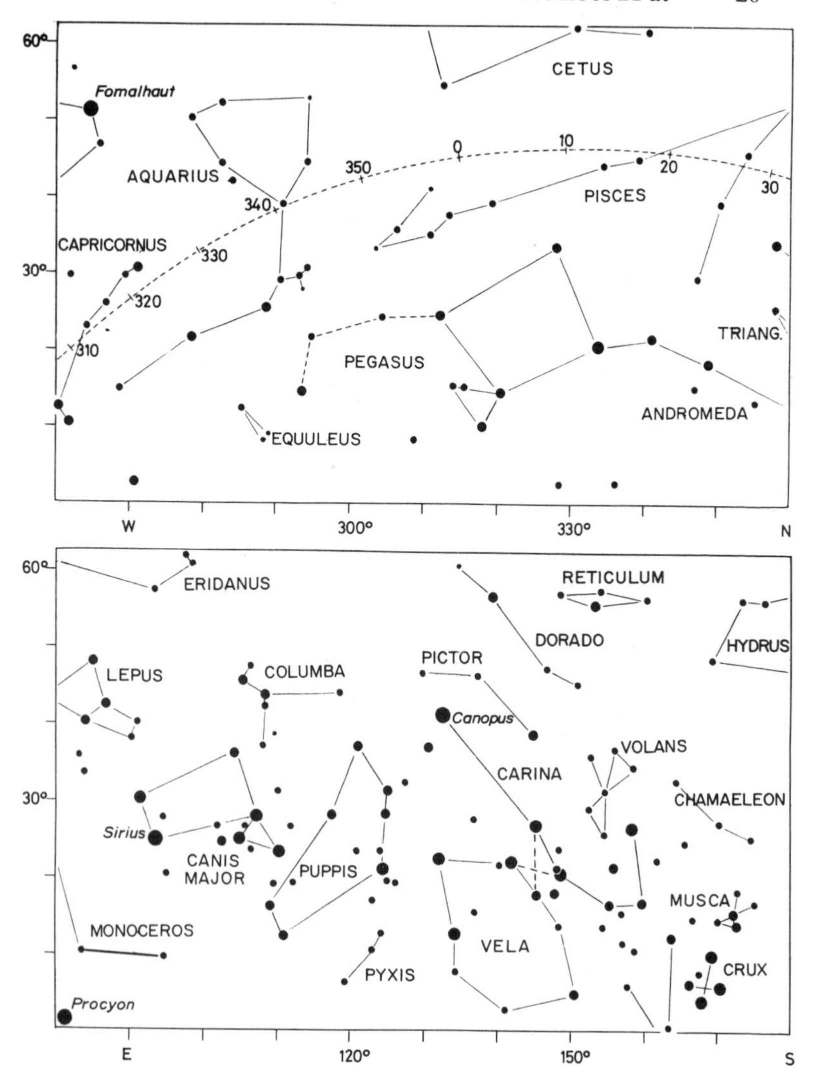

August 6 at 5^h	August 21 at 4^h	**11R**
September 6 at 3^h	September 21 at 2^h	
October 6 at 1^h	October 21 at midnight	
November 6 at 23^h	November 21 at 22^h	
December 6 at 21^h	December 21 at 20^h	

12L

September 6 at 5^h	September 21 at 4^h
October 6 at 3^h	October 21 at 2^h
November 6 at 1^h	November 21 at midnight
December 6 at 23^h	December 21 at 22^h
January 6 at 21^h	January 21 at 20^h

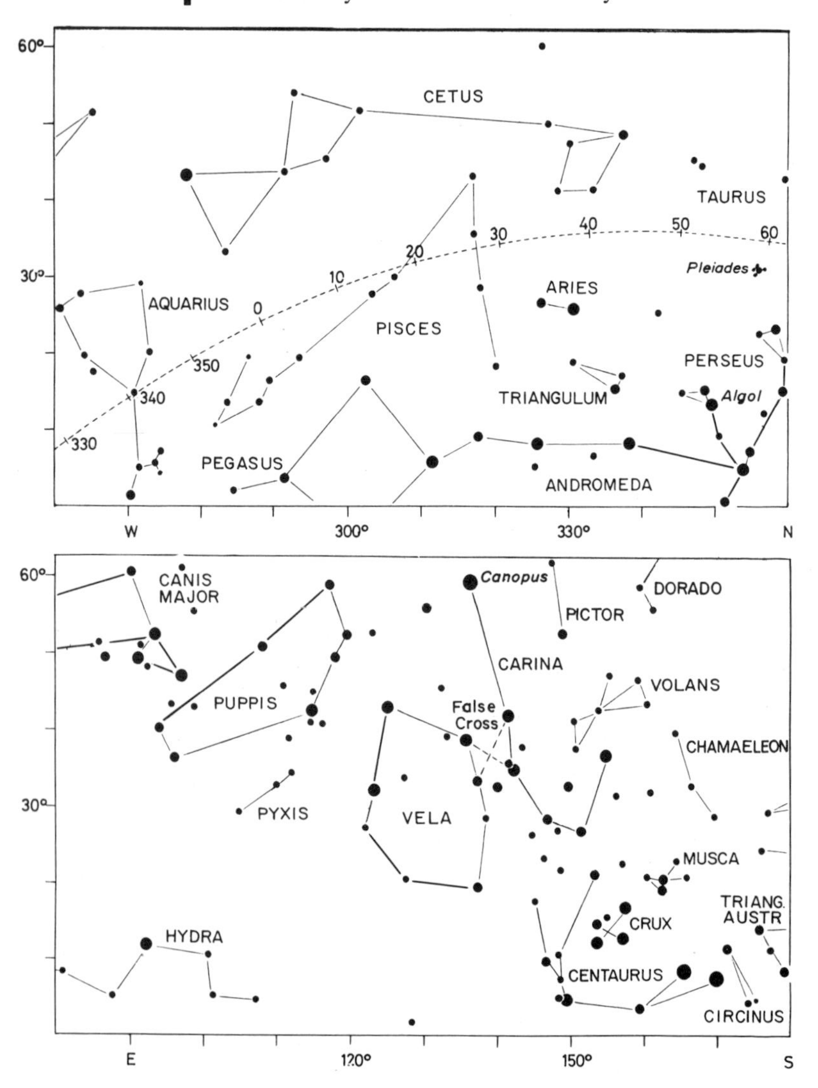

September 6 at 5h	September 21 at 4h	**12R**
October 6 at 3h	October 21 at 2h	
November 6 at 1h	November 21 at midnight	
December 6 at 23h	December 21 at 22h	
January 6 at 21h	January 21 at 20h	

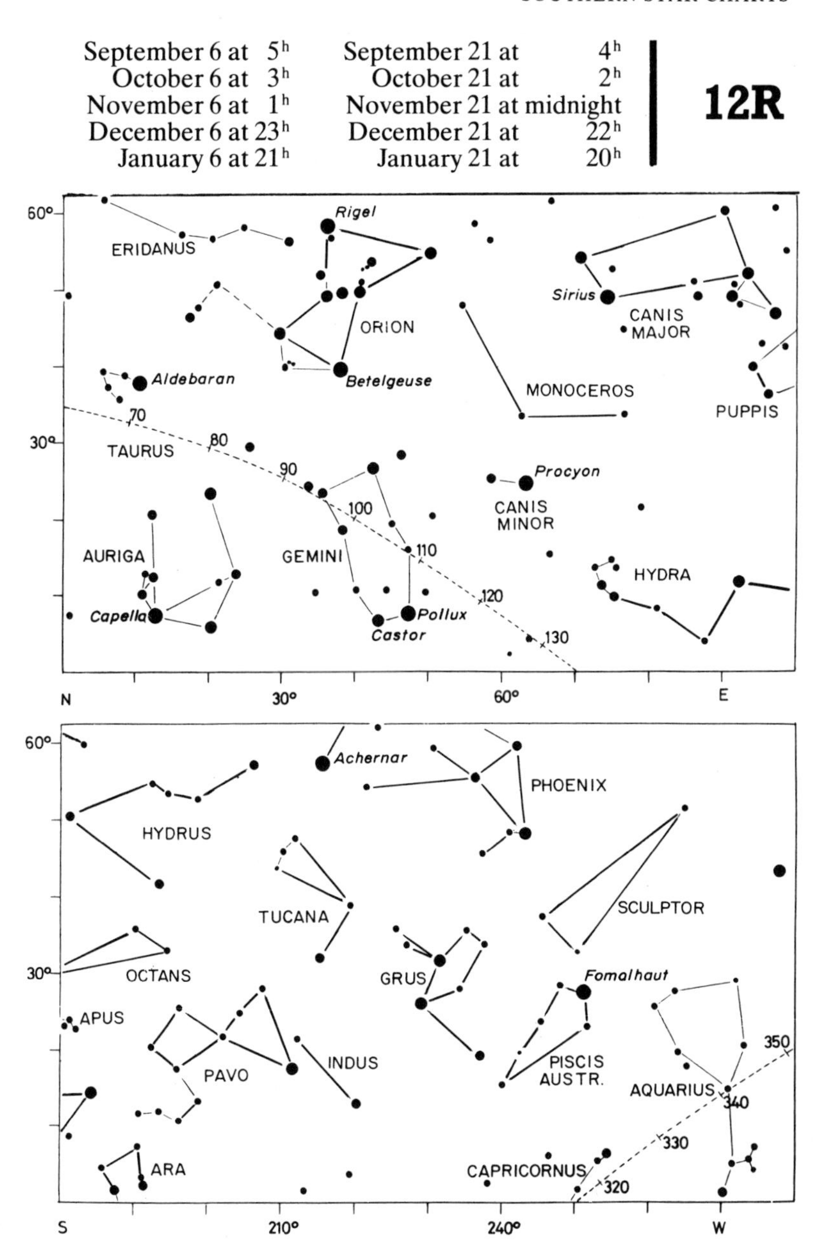

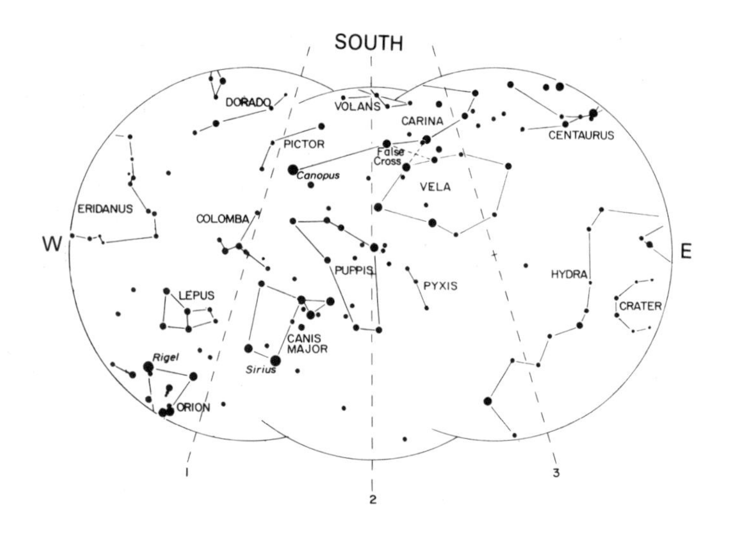

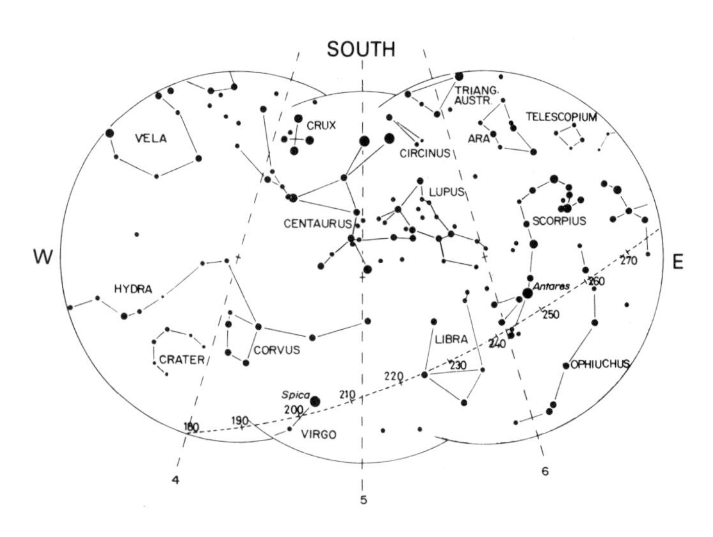

Southern Hemisphere Overhead Stars

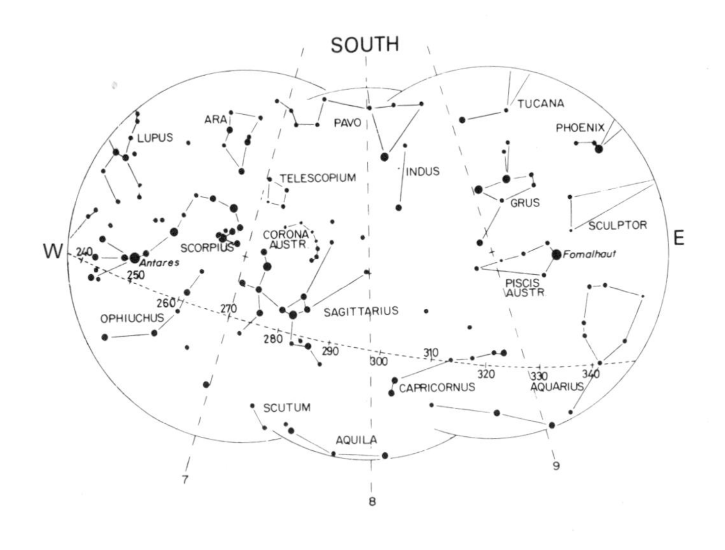

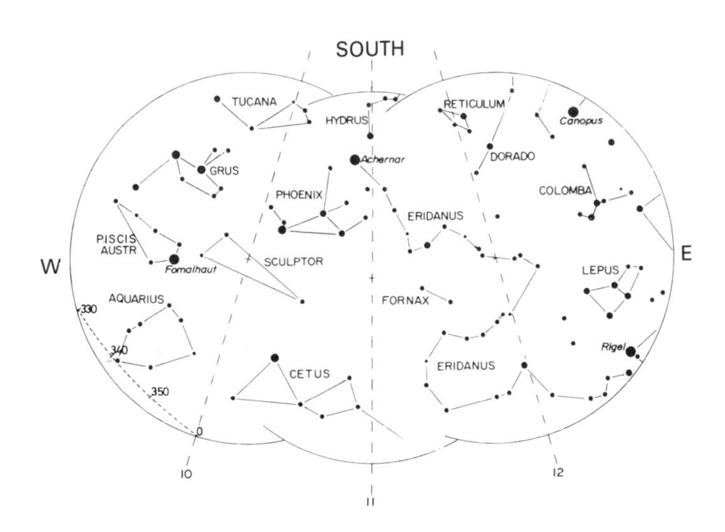

Southern Hemisphere Overhead Stars

The Planets and the Ecliptic

The paths of the planets about the Sun all lie close to the plane of the ecliptic, which is marked for us in the sky by the apparent path of the Sun among the stars, and is shown on the star charts by a broken line. The Moon and planets will always be found close to this line, never departing from it by more than about 7 degrees. Thus the planets are most favourably placed for observation when the ecliptic is well displayed, and this means that it should be as high in the sky as possible. This avoids the difficulty of finding a clear horizon, and also overcomes the problem of atmospheric absorption, which greatly reduces the light of the stars. Thus a star at an altitude of 10 degrees suffers a loss of 60 per cent of its light, which corresponds to a whole magnitude; at an altitude of only 4 degrees, the loss may amount to two magnitudes.

The position of the ecliptic in the sky is therefore of great importance, and since it is tilted at about 23½ degrees to the equator, it is only at certain times of the day or year that it is displayed to the best advantage. It will be realized that the Sun (and therefore the ecliptic) is at its highest in the sky at noon in midsummer, and at its lowest at noon in midwinter. Allowing for the daily motion of the sky, these times lead to the fact that the ecliptic is highest at midnight in winter, at sunset in the spring, at noon in summer and at sunrise in the autumn. Hence these are the best times to see the planets. Thus, if Venus is an evening star, in the western sky after sunset, it will be seen to best advantage if this occurs in the spring, when the ecliptic is high in the sky and slopes down steeply to the north-west. This means that the planet is not only higher in the sky, but will remain for a much longer period above the horizon. For similar reasons, a morning star will be seen at its best on autumn mornings before sunrise, when the ecliptic is high in the east. The outer planets, which can come to opposition (i.e. opposite the Sun), are best seen when opposition occurs in the winter months, when the ecliptic is high in the sky at midnight.

The seasons are reversed in the Southern Hemisphere, spring beginning at the September Equinox, when the Sun crosses the Equator on its way south, summer begins at the December Solstice, when the Sun is highest in the southern sky, and so on. Thus, the time when the ecliptic is highest in the sky, and therefore best placed for observing the planets, may be summarized as follows:

	Midnight	*Sunrise*	*Noon*	*Sunset*
Northern lats.	December	September	June	March
Southern lats.	June	March	December	September

In addition to the daily rotation of the celestial sphere from east to west, the planets have a motion of their own among the stars. The apparent movement is generally *direct,* i.e. to the east, in the direction of increasing longitude, but for a certain period (which depends on the distance of the planet) this apparent motion is reversed. With the outer planets this *retrograde* motion occurs about the time of opposition. Owing to the different inclination of the orbits of these planets, the actual effect is to cause the apparent path to form a loop, or sometimes an S-shaped curve. The same effect is present in the motion of the inferior planets, Mercury and Venus, but it is not so obvious, since it always occurs at the time of inferior conjunction.

The inferior planets, Mercury and Venus, move in smaller orbits than that of the Earth, and so are always seen near the Sun. They are most obvious at the times of greatest angular distance from the Sun (greatest elongation), which may reach 28 degrees for Mercury, or 47 degrees for Venus. They are then seen as evening stars in the western sky after sunset (at eastern elongations) or as morning stars in the eastern sky before sunrise (at western elongations). The succession of phenomena, conjunctions and elongations, always follows the same order, but the intervals between them are not equal. Thus, if either planet is moving round the far side of its orbit its motion will be to the east, in the same direction in which the Sun appears to be moving. It therefore takes much longer for the planet to overtake the Sun – that is, to come to superior conjunction – than it does when moving round to inferior conjunction, between Sun and Earth. The intervals given in the following table are average values; they remain fairly constant in the case of Venus, which travels in an almost circular orbit. In the case of Mercury, however, conditions vary widely because of the great eccentricity and inclination of the planet's orbit.

		Mercury	*Venus*
Inferior conj.	to Elongation West	22 days	72 days
Elongation West	to Superior conj.	36 days	220 days
Superior conj.	to Elongation East	36 days	220 days
Elongation East	to Inferior conj.	22 days	72 days

The greatest brilliancy of Venus always occurs about 36 days before or after inferior conjunction. This will be about a month *after* greatest eastern elongation (as an evening star), or a month *before* greatest western elongation (as a morning star). No such rule can be given for Mercury, because its distance from the Earth and the Sun can vary over a wide range.

Mercury is not likely to be seen unless a clear horizon is available. It is seldom seen as much as 10 degrees above the horizon in the twilight sky in northern latitudes, but this figure is often exceeded in the Southern Hemisphere. This favourable condition arises because the maximum elongation of 28 degrees can occur only when the planet is at aphelion (farthest from the Sun), and this point lies well south of the Equator. Northern observers must be content with smaller elongations, which may be as little as 18 degrees at perihelion. In general, it may be said that the most favourable times for seeing Mercury as an evening star will be in spring, some days before greatest eastern elongation; in autumn, it may be seen as a morning star some days after greatest western elongation.

Venus is the brightest of the planets and may be seen on occasions in broad daylight. Like Mercury, it is alternately a morning and an evening star, and will be highest in the sky when it is a morning star in autumn, or an evening star in spring. The phenomena of Venus given in the table above can occur only in the months of January, April, June, August and November, and it will be realized that they do not all lead to favourable apparitions of the planet. In fact, Venus is to be seen at its best as an evening star in northern latitudes when eastern elongation occurs in June. The planet is then well north of the Sun in the preceding spring months, and is a brilliant object in the evening sky over a long period. In the Southern Hemisphere a November elongation is best. For similar reasons, Venus gives a prolonged display as a morning star in the months following western elongation in November (in northern latitudes) or in June (in the Southern Hemisphere).

The superior planets, which travel in orbits larger than that of the Earth, differ from Mercury and Venus in that they can be seen opposite the Sun in the sky. The superior planets are morning stars after conjunction with the Sun, rising earlier each day until they come to opposition. They will then be nearest to the Earth (and therefore at their brightest), and will be on the meridian at midnight, due south in northern latitudes, but due north in the Southern Hemisphere. After opposition they are evening stars, setting earlier each evening until they set in the west with the Sun at the next conjunction. The change in brightness about the time of opposition is most noticeable in the case of Mars, whose distance from the Earth can vary considerably and rapidly. The other superior planets are at such great distances that there is very little change in brightness from one opposition to another. The effect of altitude is, however, of some importance, for at a December opposition in northern latitudes the planet will be among the stars of Taurus or Gemini, and can then be at an altitude of more than 60 degrees in southern England. At a summer opposition, when the planet is in Sagittarius, it may only rise to about 15 degrees above the southern horizon, and so makes a less impressive appearance. In the Southern Hemisphere, the reverse conditions apply; a June opposition being the best, with the planet in Sagittarius at an altitude which can reach 78 degrees above the northern horizon.

Mars, whose orbit is appreciably eccentric, comes nearest to the Earth at an opposition at the end of August. It may then be brighter even than Jupiter, but rather low in the sky in Aquarius for northern observers, though very well placed for those in southern latitudes. These favourable oppositions occur every fifteen or seventeen years (1956, 1971, 1988, 2003) but in the Northern Hemisphere the planet is probably better seen at an opposition in the autumn or winter months, when it is higher in the sky. Oppositions of Mars occur at an average interval of 780 days, and during this time the planet makes a complete circuit of the sky.

Jupiter is always a bright planet, and comes to opposition a month later each year, having moved, roughly speaking, from one Zodiacal constellation to the next.

Saturn moves much more slowly than Jupiter, and may remain in the same constellation for several years. The brightness of Saturn depends on the aspect of its rings, as well as on the distance from Earth and Sun. The rings are now inclined towards the Earth and Sun at quite a small angle, and are opening again after being seen edge-on in 1980. The next passage of both the Earth and the Sun through the ring-plane will not occur until 1995.

Uranus, Neptune, and *Pluto* are hardly likely to attract the attention of observers without adequate instruments, but some notes on their present positions in the sky will be found in the April and June Notes.

Phases of the Moon 1984

New Moon				*First Quarter*				*Full Moon*				*Last Quarter*			
	d	h	m		d	h	m		d	h	m		d	h	m
Jan.	3	05	16	Jan.	11	09	48	Jan.	18	14	05	Jan.	25	04	48
Feb.	1	23	46	Feb.	10	04	00	Feb.	17	00	41	Feb.	23	17	12
Mar.	2	18	31	Mar.	10	18	27	Mar.	17	10	10	Mar.	24	07	58
Apr.	1	12	10	Apr.	9	04	51	Apr.	15	19	11	Apr	23	00	26
May	1	03	45	May	8	11	50	May	15	04	29	May	22	17	45
May	30	16	48	June	6	16	42	June	13	14	42	June	21	11	10
June	29	03	18	July	5	21	04	July	13	02	20	July	21	04	01
July	28	11	51	Aug.	4	02	33	Aug.	11	15	43	Aug.	19	19	40
Aug.	26	19	25	Sept.	2	10	30	Sept.	10	07	01	Sept.	18	09	31
Sept.	25	03	11	Oct.	1	21	52	Oct.	9	23	58	Oct.	17	21	14
Oct.	24	12	08	Oct.	31	13	07	Nov.	8	17	43	Nov.	16	06	59
Nov.	22	22	57	Nov.	30	08	00	Dec.	8	10	53	Dec.	15	15	25
Dec.	22	11	47	Dec.	30	05	27								

All times are G.M.T.

Reproduced, with permission, from data supplied by
the Science and Engineering Research Council

Longitudes of the Sun, Moon and Planets in 1984

DATE		Sun °	Moon °	Venus °	Mars °	Jupiter °	Saturn °
January	6	284	315	245	207	267	224
	21	300	154	264	215	270	225
February	6	316	359	283	222	273	226
	21	331	206	302	228	276	226
March	6	345	20	319	233	278	226
	21	0	229	338	236	280	225
April	6	16	67	357	238	282	225
	21	31	277	16	236	282	224
May	6	45	103	34	232	282	222
	21	60	310	53	227	282	221
June	6	75	156	73	223	280	220
	21	89	354	91	221	279	220
July	6	104	195	109	223	277	219
	21	118	26	128	227	275	219
August	6	133	247	147	234	274	220
	21	148	72	166	241	273	220
September	6	163	295	185	250	273	222
	21	178	120	204	260	273	223
October	6	192	329	222	270	275	224
	21	207	158	240	281	277	226
November	6	223	13	260	292	279	228
	21	238	212	278	303	282	230
December	6	254	45	296	315	285	232
	21	269	250	313	326	288	233

Longitudes of *Uranus* 250° *Neptune* 270° (at opposition)

Moon: Longitude of ascending node
Jan. 1: 75° Dec. 31: 55°

Mercury moves so quickly among the stars that it is not possible to indicate its position on the star charts at a convenient interval. The monthly notes must be consulted for the best times at which the planet may be seen.

The positions of the other planets are given in the table on the previous page. This gives the apparent longitudes on dates which correspond to those of the star charts, and the position of the planet may at once be found near the ecliptic at the given longitude.

Examples:

In the southern hemisphere two planets are seen in the western evening sky in mid-October. Identify them.

The southern star chart 9L shows the western sky at Oct. $6^{d}21^{h}$ (or mid-October at 20^{h}) and shows longitudes 260°–300°. Reference to the table opposite gives the longitudes of both Mars and Jupiter as 276°. Thus these planets are found in Sagittarius and the one with the slightly reddish tint is Mars.

The positions of the Sun and Moon can be plotted on the star maps in the same manner as for the planets. The average daily motion of the Sun is 1°, and of the Moon 13°. For the Moon an indication of its position relative to the ecliptic may be obtained from a consideration of its longitude relative to that of the ascending node. The latter changes only slowly during the year as will be seen from the values given on the opposite page. Let us call the difference in longitude of Moon-node, d. Then if d = 0°, 180° or 360° the Moon is on the ecliptic. If d = 90° the Moon is 5° north of the ecliptic and if d = 270° the Moon is 5° south of the ecliptic.

On October 6 the Moon's longitude is given as 329° and the longitude of the node is found by interpolation to be about 60°. Thus d = 269° and the Moon is about 5° south of the ecliptic. Its position may be plotted on northern star charts 8L, 10R and 11R and southern star charts 10L, 11R and 12R.

Some Events in 1984

ECLIPSES

In 1984 there will be two eclipses, both of the Sun.

30 May: annular eclipse of the Sun – North America, Arctic regions, Greenland, Iceland, north-west of South America, West Africa, Europe.

22–23 November: total eclipse of the Sun – Indonesia, Australia, Papua New Guinea, Antarctica, New Zealand, South America.

THE PLANETS

Mercury may be seen more easily from northern latitudes in the evenings about the time of greatest eastern elongation (3 April) and in the mornings around greatest western elongation (September 14). In the Southern Hemisphere the corresponding most favourable dates are around 22 January (mornings) and 1 August (evenings).

Venus is visible in the mornings until the end of April. After superior conjunction it is visible in the evenings from August onwards.

Mars is at opposition on 11 May.

Jupiter is at opposition on 29 June.

Saturn is at opposition on 3 May.

Uranus is at opposition on 1 June.

Neptune is at opposition on 21 June.

Pluto is at opposition on 20 April.

January

New Moon: 3 January *Full Moon:* 18 January

Earth is at perihelion (nearest to the Sun) on 3 January at a distance of 147 million kilometres.

Mercury attains its greatest western elongation (24°) on 22 January. It is further south than the Sun and thus poorly placed for observation by those in northern temperate latitudes, where it may only be seen during the middle ten days of January low above the south-eastern horizon, about the time of beginning of morning civil twilight. However, Mercury is much better situated for observation by those in the Southern Hemisphere where the planet will best be seen above the east-south-eastern horizon about half-an-hour before sunrise, after the end of the first week of the month. Mercury increases in brightness quite markedly during this period, magnitude from +1½ to 0. Do not confuse Mercury with the much brighter planet Jupiter. At first the two planets are only a few degrees apart but Jupiter gradually draws away from both the Sun and Mercury.

Venus, like Mercury, is also visible in the morning skies, though much further from the Sun than the latter. The magnitude of Venus is −3.5. Venus passes only 1° north of Jupiter on 27 January.

Mars is visible as a morning object, magnitude +1.2. It begins the month in Virgo and then moves eastwards into Libra. Mars reaches opposition in May and is visible at some time of night throughout the whole of 1984. On 25 January the Last Quarter Moon passes north of Mars, passing Saturn a day later and Jupiter on 29 January.

Jupiter, magnitude −1.4, is visible as a morning object, above the south-eastern horizon before dawn. Because of its southern declination observers in northern temperate latitudes will not be able to see it before the middle of the month. For such observers it will never be very high above the horizon, even when on the meridian, since Jupiter is in Sagittarius throughout 1984.

Saturn remains in the constellation of Libra throughout the year and begins by being visible as a morning object, magnitude +0.8, in the south-eastern sky before the twilight gets too strong. The Moon, near Last Quarter, passes Saturn early on 26 January, an actual occultation being visible from Australia.

THE ECLIPSE OF EPSILON AURIGÆ

The eclipsing binary Epsilon Aurigæ, close to Capella in the sky and therefore excellently placed for observation during winter evenings in the northern hemisphere, is a remarkable system. It was described in the 1983 *Yearbook* (pages 86–7), when its main characteristics were discussed. The eclipse, which began on 22 July 1982, became total on 11 January 1983, by which time the fading had become very noticeable; instead of being brighter than its neighbour Eta Aurigæ, Epsilon was at least half a magnitude fainter. Totality ends on 16 January 1984, and from then until 25 June, when the eclipse is over, Epsilon Aurigæ will slowly recover its lost light, after which it will fluctuate no more until the start of the next eclipse in the year 2011. Much attention has been paid to the system recently, though we are still unclear as to the precise nature of the invisible eclipsing secondary.

ETA GEMINORUM

Eta Geminorum (sometimes called by its old name of Propus) is also a variable star and a member of an eclipsing system, but it is very different from Epsilon Aurigæ. It is usually said to be a semi-regular variable with a mean period of 233 days and an amplitude of from magnitude 3.1 to 3.9, but this is an over-simplification.

The primary is a red giant of spectral type M3; its distance is 185 light-years, and its mean luminosity is about 130 times that of the Sun. There is an optical companion of magnitude 6.5, discovered by S. W. Burnham in 1881; the present separation is about 1°.5 and the position angle 266°. The spectral type is G8, and the luminosity is about 7 times that of the Sun.

However, the main star is itself a spectroscopic binary as well as being intrinsically variable (the variability was discovered by Julius

Schmidt, one of this year's centenaries, in 1844). The secondary may also be of type M, but there is no definite information. Eclipses also are uncertain in their timing, though one certainly occurred in the winter of 1979–80. At times when eclipse coincides with minimum magnitude, the star may decline to around 3.7, but there seems no certain evidence that it fades as low as 3.9, and the general value is between 3.1 and 3.4. Neither is the 233-day period at all well established. This makes Eta Geminorum a good subject for naked-eye observation; convenient comparison stars are Epsilon Geminorum (2.98) and Xi (3.36). Mu Geminorum, also an M-type star lying near Eta, should be avoided inasmuch as it is itself slightly variable – with an official magnitude of 2.88.

ERNST FRIEDRICH WILHELM KLINKERFUES

This month's centenary is that of the death of a leading German astronomer. Klinkerfues was born at Hofgeisnau, in Hesse-Cassel, on 29 March 1827, and was educated at Göttingen, becoming Director of the Observatory there in 1859 and Professor of Astronomy at the University in 1861. He was concerned with the measurements of stellar parallaxes and the orbits of binary stars, but his main work was in connection with comets. In particular, he was concerned with an extraordinary episode concerning Biela's Comet, which had been a regular visitor with a period of 6¾ years. At its return of 1845 it split into two pieces. The twins returned in 1852, rather more widely separated; were missed in 1859, though their position in the sky was unfavourable; and failed to put in an appearance either at the expected return of 1866 or subsequently. Most astronomers regarded the comet as lost, but Klinkerfues continued to pay attention to the subject, and on 30 November 1872 he sent a telegram to the well-known observer Pogson, who was at Madras: 'Biela touched Earth on 27th. Search near Theta Centauri.' Pogson did so – and observed a comet on 2 and 3 December, but was unable to follow it further because of bad weather, and nobody else managed to see it at all.

There seems no doubt that the comet seen by Pogson was not Biela's, though a shower of meteors was observed both in 1872 and at other years when the comet should have been back; a few meteors of the shower are still seen every November, though the rate is now very feeble. The mystery of what Klinkerfues really meant, and what Pogson's comet was, remains unsolved.

THE QUADRANTIDS

This is a very brief shower, but can be very spectacular (see the article by John Mason in the 1981 *Yearbook*). Conditions in 1984 will be favourable, as on the night of 3–4 January there will be no interference from the Moon.

February

New Moon: 1 February *Full Moon:* 17 February

Mercury continues to be well placed for observation by those in the Southern Hemisphere for the first three weeks of the month. It is still best seen above the east-north-east horizon about half-an-hour before sunrise. By the end of this period of visibility Mercury has increased in brightness to magnitude −0.5. For observers in northern temperate latitudes the planet remains unsuitably placed for observation.

Venus is a brilliant object in the morning skies before sunrise, magnitude −3.4. Being well south of the equator the planet will only be seen by observers in northern temperate latitudes for a short period before sunrise, low in the south-eastern sky.

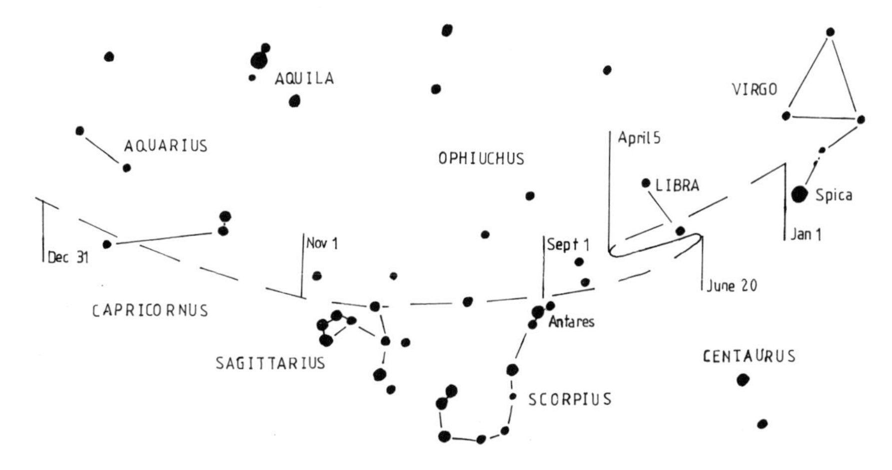

Figure 1. Path of Mars.

Mars continues to be visible in the mornings, and actually brightens by about half a magnitude during the month. Its path amongst the stars during the year is shown in Figure 1. When Mars passes within 1° of Saturn on 15 February the two planets will be of equal brightness, Mars being distinguishable because of its slightly reddish tinge. The gibbous Moon will be close to both planets on 22 February.

Jupiter, magnitude −1.5, is visible as a morning object in the south-eastern sky.

Saturn, magnitude +0.7, remains visible in the south-eastern sky in the early mornings. Its path amongst the stars is shown in Figure 4 (given with the notes for April).

SCHMIDT AND LINNÉ

A hundred years ago, on 8 February 1884, Johann Friedrich Julius Schmidt – always remembered as Julius Schmidt – died in Athens. He was a distinguished astronomer who discovered the bright nova Q Cygni in 1876, but will always be remembered for his association with the alleged change in the lunar formation Linné.

Schmidt was German, born at Eutin in Lübeck on 25 October 1825. At the age of fourteen he began to observe the Moon, and determined to draw up a large map of the lunar surface. At that period the best map was that of Beer and Mädler, published in 1838–9; it was a masterpiece of careful, painstaking work, but it had been compiled with a small telescope (Beer's 3¾-inch Fraunhofer refractor at his private observatory near Berlin), and obviously it was limited. A more elaborate chart had been started by Lohrmann of Dresden, but because of ill-health Lohrmann was unable to complete it.

Meantime, Schmidt had completed his education at Hamburg, and had then gone to Altona Observatory, and thence, in 1853, to Olmütz. In 1858 he was appointed Director of the Athens Observatory, and went to Greece, where he remained for the rest of his life. It was here, in 1866, that he made a startling claim. On the lunar Mare Serenitatis (Sea of Serenity) Beer and Mädler had drawn a small but well-marked crater, and named it Linné in honour of the great Swedish scientist Linnæus. Schmidt announced the crater had disappeared, to be replaced by nothing more than a white patch. Subsequently a small craterlet was recorded in the middle of this patch.

The effect of Schmidt's announcement was nothing short of dramatic. Beer and Mädler had regarded the Moon as changeless, and, ironically, their map (plus their book, giving a detailed description of each named formation) had actually diverted attention from the Moon; if there were no variations, astronomers in general saw little point in observing it further. However, the Linné episode caused attention to be directed back to lunar work. Many authorities were confident that a change had occurred; for instance Sir John Herschel commented that the most likely cause was a collapse of the crater-walls.

The controversy went on for many years, but a close examination of the evidence shows that any real alteration must be dismissed as extremely unlikely Few observers were paying much attention to the Moon between 1840 and 1866, and indeed Mädler stated in 1868 that Linné looked exactly the same as he remembered seeing it in the 1830s. Orbiter pictures of Linné taken from close range show a well-defined crater which has a light surround; there is no indication that any change has occurred there for many millions of years. Yet the whole episode was beneficial; since 1866 the Moon has been kept under constant survey.

Schmidt continued his work. In 1878 he completed his large lunar map; he had made use of the original sections by Lohrmann, but the major work was Schmidt's own, and his map remained the best for many years – a great tribute to the skill, energy and persistence of its compiler.

March

New Moon: 2 March *Full Moon:* 17 March

Summer Time in Great Britain and Northern Ireland commences on 25 March.

Equinox: 20 March

Mercury passes through superior conjunction on 8 March and a fortnight later begins to be visible as an evening object, reaching greatest eastern elongation (19°) on 3 April. Being well north of the Sun it is observers in the Northern Hemisphere who will get the best view. Indeed, this evening apparition, which extends through the first two weeks of April is the most suitable one of the year for observers in northern temperate latitudes. Figure 2 shows

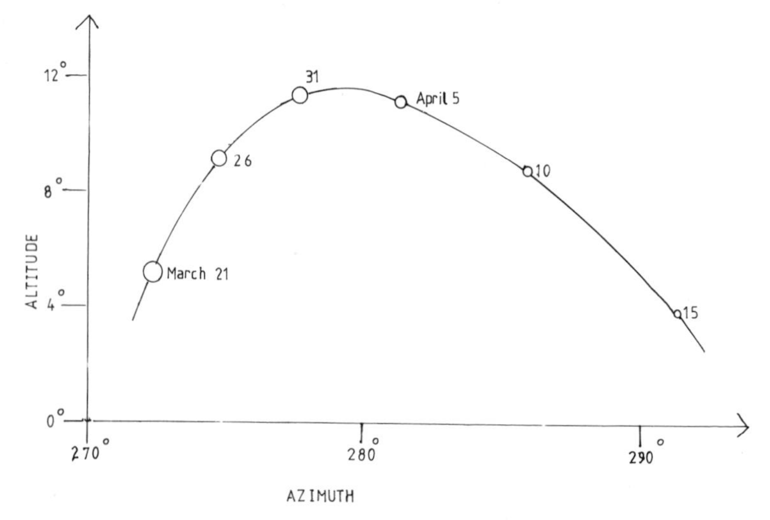

Figure 2. Mercury for N.52°

changes in azimuth (true bearing from the north through east, south, and west) and altitude of Mercury on successive evenings when the Sun is 6° below the horizon in latitude N.52°. This condition is known as the end of evening civil twilight, and in this latitude and at this time of year occurs about 40 minutes after sunset. The changes in brightness of the planet are roughly indicated by the sizes of the circles marking positions at 5-day intervals and it will be seen that Mercury is brightest before it reaches eastern elongation. Its magnitude on 21 March is −1.2 but this has fallen to +2.1 by 15 April.

Venus continues to be visible as a brilliant morning object. It is slowly drawing in towards the Sun and observers in northern temperate latitudes will be finding it increasingly difficult to observe and by the end of March Venus will be rising at about the same time as the Sun.

Mars is still in Libra and is well placed for observation near the meridian before dawn. The passage of the gibbous Moon past Mars on 21 March gives rise to a grazing occultation visible from the northern tip of North Island, New Zealand.

Jupiter, magnitude −1.6, is visible as a morning object for several hours before sunrise, in the south-eastern sky.

Saturn continues to be visible as a morning object, magnitude +0.6, in the south-eastern sky in the early morning. An occultation of the planet by the gibbous Moon on 20 March is visible from Australia and New Zealand.

OCCULTATION OF A STAR BY SATURN'S RINGS

Observable occultations of stars by the rings of Saturn are rare events since the brightness of the rings renders faint stars invisible before the acutal occultation occurs. Even with modern equipment the limiting magnitude for an occultation to be successfully observed is probably about +6 to +9, depending on the spectral type of the star concerned.

The first useful observation of an occultation of a star by Saturn's rings occurred in 1917 and was witnessed by three unprepared observers in England namely Ainslie, Knight, and Mrs Freeman, who observed the star BD+21° 1714 pass behind the rings. The observation demonstrated the transparency of Ring A, the freedom of the Cassini division from obscuring matter, and the partial transparency of the Encke division. Ainslie also deduced that the particles in Ring A were considerably smaller than 6 km in diameter and Crommelin suggested that 0.2 km was more likely for the upper limit if the star's angular diameter was 0″.0001 rather than the figure of 0″.001 used by Ainslie.

The circumstances of the occultation of the star SAO 158913 ($8^m.8$) which occurs on 25 March are illustrated in the diagram. (Figure 3). The area of visibility is America, the Pacific Ocean, and New Zealand. The predicted times are given below.

	Station	*Disappearance* U.T.	P.	*Alt. of Star*	*Alt. of Sun*	*Reappearance* U.T.	P.	*Alt. of Star*	*Alt. of Sun*
		h m	°	°	°	*h m*	°	°	°
Saturn	Washington	6 38	305	34	−45	9 01	95	35	−24
	Fort Davis	6 40	303	28	−57	9 04	96	45	−46
	La Silla	6 33	298	70	−53	8 59	100	64	−25
	Hawaii	(below horizon)				9 10	99	22	−60
	Wellington	(below horizon)				9 07	105	9	−30
Saturn's Rings (outer edge of Ring A)	Washington	6 07	300	32	−47	10 18	102	27	−10
	Fort Davis	6 05	298	22	−55	10 19	103	44	−33
	La Silla	5 51	295	63	−59	10 07	105	50	−10
	Hawaii	(below horizon)				10 21	104	36	−67
	Wellington	(below horizon)				10 10	107	21	−40
Saturn's Rings (inner edge of Ring C)	Washington	(behind planet)				9 16	97	34	−21
	Fort Davis	(behind planet)				9 18	98	45	−44
	La Silla	(behind planet)				9 08	102	62	−23
	Hawaii	(behind planet)				9 20	100	24	−62
	Wellington	(behind planet)				9 10	105	10	−31

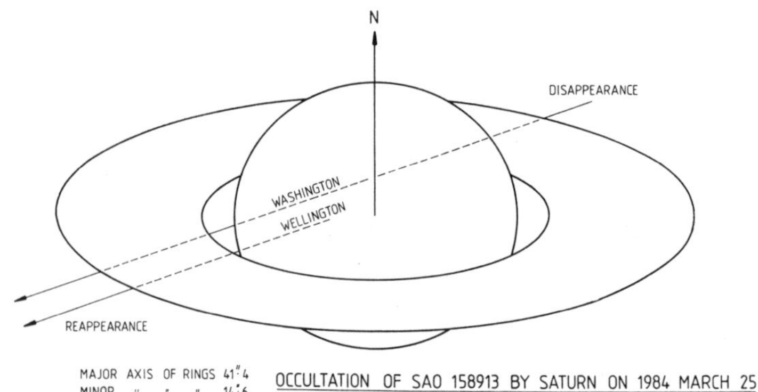

Figure 3

JOHN HERSCHEL AND α HYDRÆ

Alphard, or α Hydræ, is well placed during March evenings, and since it is less than 9° south of the celestial equator it is visible from every inhabited continent. It has a K3-type spectrum; its distance is 85 light-years, and it is more than 700 times as luminous as the Sun.

In early 1838 Sir John Herschel was sailing home to England from the Cape, having completed his survey of the southern stars. During the voyage he made several naked-eye observations of Alphard, which he regarded as definitely variable.

21 March. α Hydræ inferior to δ Canis Majoris, brighter than δ Argûs (now known as δ Velorum) and γ Leonis.

7 May. α Hydræ fainter than β Aurigæ, very obviously fainter than γ Leonis, Polaris, or β Ursæ Minoris. 'Though low, yet it is now a decidedly insignificant star,' wrote Herschel. 'Moon, and low altitude, but it leaves no doubt in my mind of the minimum being nearly attained.'

8 May. 'α Hydræ still high (30°) and quite free of all cloud and haze in a very fine blue sky . . . is very decidedly inferior to γ Leonis. . . . I incline to place the minimum of last night as α is tonight brighter than β Aurigæ at certain intervals of its twinkling. . . . On the whole after many comparisons α Hydræ is rather inferior to β Aurigæ.'

9 May. 'α Hydræ inferior to γ Leonis but . . . evidently on the rapid increase. . . . The minimum is certainly fairly passed and the star is rapidly regaining its light.'

10 May. 'α Hydræ much inferior to γ Leonis, rather inferior to β Aurigæ. It is still about its minimum.'

11 May. 'α Hydræ brighter than β Aurigæ no doubt; β much higher.'

12 May. 'Castor and α Hydræ nearly equal.'

Herschel docked in London on 15 May, and apparently made no further observations of the star.

Modern magnitude measures make α Hydræ 1.98, β Aurigæ 1.90, δ Velorum 1.96, β Ursæ Minoris 2.08, Polaris 1.99 (very slightly variable), γ Leonis 1.99 (the combined magnitude of its components) and Castor 1.58 (again a combined magnitude). α Hydræ is an awkward star to estimate because it is so isolated, but it is probably worth watching, though whether or not it is perceptibly variable seems dubious.

April

New Moon: 1 April *Full Moon:* 15 April

Mercury continues to be visible in the evenings for the first half of the month, though not to observers in southern latitudes. Northern Hemisphere observers should refer to the diagram given with the March notes.

Venus, already lost to view for observers in northern temperate latitudes, continues to move in towards the Sun and for observers further south is only visible for a short while before dawn low above the eastern horizon.

Mars is now a conspicuous object in the morning skies as its magnitude increases from −0.5 to −1.4 during the month. Mars reaches its first stationary point on 5 April and then moves retrograde, in Libra.

Jupiter, magnitude −1.9, continues to be visible in Sagittarius as a morning object. Observers with a good pair of binoculars, providing that they are steadily supported, should attempt to detect the four Galilean satellites. The main difficulty in observing them is the overpowering brightness of Jupiter itself. The path of Jupiter amongst the stars is shown in Figure 6 (see the notes for June).

Saturn is slowly increasing in brightness (to magnitude +0.4) as it moves towards opposition early next month. Saturn is in Libra (see Figure 4).

Pluto is at opposition on 20 April in the north-eastern part of Virgo. The planet is not visible to the naked-eye, since its magnitude is +14. When closest to the Earth its distance is 4,318 million kilometres.

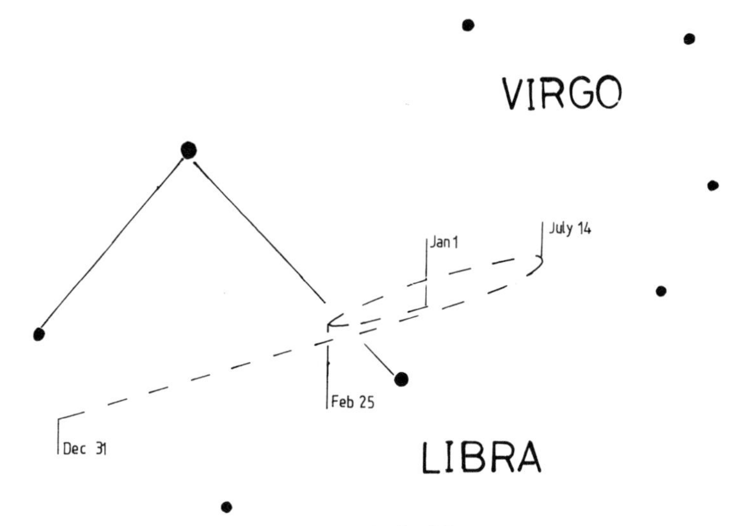

Figure 4. Path of Saturn.

THE FURTHEST PLANET

In almost every astronomical text, the most distant planets are listed as Uranus, Neptune, and Pluto, in this order. Reckoning according to their mean distances from the Sun, this is true enough. However, Pluto's eccentric orbit means that at perihelion it is closer-in than Neptune. Perihelion is due in 1989; between 1979 and 1999 the distance of Pluto from the Sun is less than that of Neptune. In any case, Pluto's status as a planet is now seriously in doubt, and there have been suggestions that it and its companion, Charon, should more properly be regarded as a double asteroid.

PIONEER 10

Pioneer 10, launched in 1972, is already more remote than either planet. It crossed the orbit of Pluto on 25 April 1983, when it was 2,779,209,908 miles (4,472,497,438 km) from the Sun, moving at 30,613 m.p.h. Neptune's orbit was crossed on 13 June 1983, at 2,813,685,909 miles (4,527,978,612 km), the speed was then 30,558 m.p.h. It is still working well; only its magnetometer has failed, and it is sending back data about the solar heliosphere, which has proved to be larger than expected, and appears to 'breathe' in and out according to the 11-year solar cycle. Also, the main source of turbulence in the outer heliosphere is due to storms on the Sun.

Near solar maximum, cosmic rays from beyond the Solar System are largely blocked by the heliosphere, and as solar storm activity builds up, the heliosphere changes from a reasonably circular shape into an oval.

After passing through the Oort comet cloud, at around 50,000 astronomical units from the Sun, Pioneer 10 will leave the Solar System permanently. The first star to be approached (to within 3.3 light-years) will be Ross 248, in 32,610 years' time. No other star will be encountered for a further 800,000 years, so that Pioneer has a long journey ahead of it. Presumably it will continue on its way until it is destroyed by collision with a large body, or else (much less probably!) collected by some alien civilization, who will do their best to interpret the plaque which it carries.

CRATER

Crater, the Cup, is one of the original 48 constellations listed by Ptolemy, but it is extremely obscure, and none too easy to identify. It has seven stars above the fifth magnitude. They are:

Star	*Magnitude*	*Spectrum*	*Distance (light-years)*	*Luminosity (Sun=1)*
α (Alkes)	4.08	K0	121	60
β	4.48	A2	235	60
δ	3.56	G8	72	16
γ	4.08	A5	78	10
θ	4.70	B9	257	60
ε	4.83	K5	310	130
ζ	4.73	G8	240	55

The brightest star is therefore not Alkes, as would be expected in view of the fact that it is lettered α, but δ, which is half a magnitude brighter. There is little of telescopic interest in the constellation, though telescopic users will be interested to locate the very red variable R Crateris, which lies near Alkes.

CHARLES OLIVIER

10 April is the centenary of the birth of Charles Pollard Olivier, a leading authority on meteors. He was born at Charlottesville in Virginia, and was educated at Virginia University, where he became Professor of Astronomy; in 1928 he went to the University of Pennsylvania as Professor and Director of the Flower Observatory. He catalogued 1,200 separate meteor streams, and was the author of a standard work on the subject, published 1925.

May

New Moons: 1 May, 30 May *Full Moon:* 15 May

Mercury attains its greatest western elongation (26°) on 19 May and is therefore visible as a morning object though only for observers in equatorial regions and the Southern Hemisphere. Figure 5 shows the changes in azimuth and altitude of Mercury on successive mornings when the Sun is 6° below the horizon. This condition, known as the beginning of morning civil twilight occurs about 30 minutes before sunrise in latitude S.35°, at this time of year. This apparition extends for about six weeks and the increase in brightness of the planet during this period is quite marked, amounting to nearly three magnitudes.

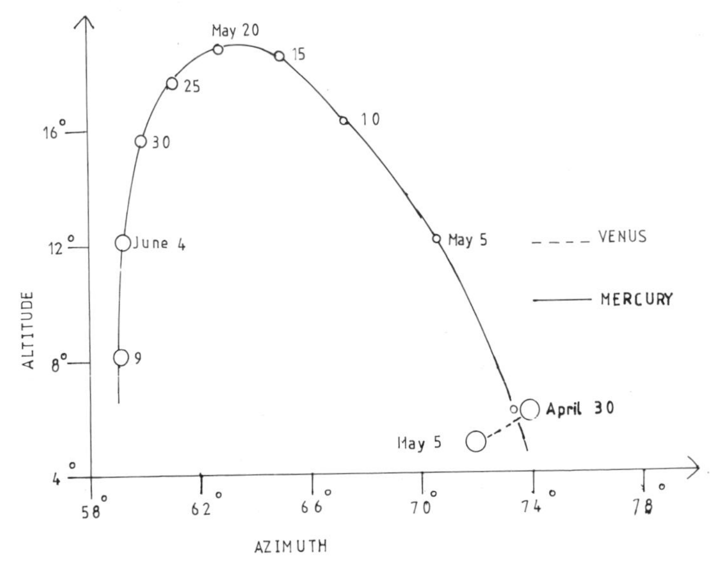

Figure 5. Mercury and Venus for S.35°.

Venus soon becomes lost in the morning twilight as it moves in towards superior conjunction next month. However, its proximity to Mercury as it emerges from the morning twilight at the very beginning of the month is shown in Figure 5, though the long duration of twilight inhibits observation from northern temperate latitudes.

Mars reaches opposition on 11 May, magnitude −1.7, and thus is available for observation throughout the night. The eccentricity of the orbit of the planet is such that its closest approach to the Earth (80 million kilometres) does not occur until 19 May, eight days after opposition (this is almost the maximum difference). The Full Moon passes just north of Mars on the early evening of 19 May, having passed Saturn only half-a-day earlier. It is of interest to note that Mars and the Sun are so diametrically opposite to each other in the sky, as seen from the Earth, that an observer on Mars would actually be able to observe the Earth in transit across the face of the Sun on 11 May – an occurrence which will not be repeated until A.D. 2084.

Jupiter continues to be a conspicuous object in the mornings, magnitude −2.0.

Saturn is visible throughout the night during the month since it is at opposition on 3 May. Saturn has a magnitude of +0.3 and when at opposition is 1,324 million kilometres from the Earth.

An annular eclipse of the Sun will occur on 30 May; see following note and page 121.

OTHER SOLAR ECLIPSES

The track of the annular eclipse on 30 May extends across various populated countries, including Mexico, and will be widely observed. However, there is no hope of seeing the corona or the prominences; for this, the eclipse must be completely total.

The fact that the Sun and Moon appear so nearly equal in size in our sky must be put down to coincidence. It is a fortunate one, since otherwise we would know much less about the Sun's surroundings than we actually do (remember, it was only in the mid-nineteenth century that all astronomers were satisfied that the prominences belong to the Sun rather than the Moon!).

There is no other planet which has a satellite which can 'just' produce a total eclipse as seen from the planet itself. For instance, to a Martian observer both Phobos and Deimos would appear much too small, though they would transit the Sun frequently. Phobos would do so 1,300 times in each Martian year, taking 19 seconds to cross the disk, while Deimos would transit 130 times on average, taking 1 minute 48 seconds to cross the Sun. From Jupiter, the mean diameter of the Sun is less than 6 minutes of arc, and total eclipses could be produced by all the Galilean satellites as well as Amalthea; from the surface of Jupiter Io would subtend an angular diameter of over 35 minutes of arc. As seen from Saturn, the diameter of the Sun is only 3′ 22″, so that total eclipses can be produced by all the principal satellites apart from Hyperion and Iapetus; the largest satellite as seen from Saturn is Tethys, with a diameter of over 17 minutes of arc. From Uranus the Sun's diameter is reduced to below 2 minutes of arc, and all five satellites can produce total eclipses; the largest would be Ariel, diameter almost 31 minutes of arc. From Neptune the Sun's diameter is only about 1 minute of arc, and Triton has an apparent diameter of over one degree, though it must be remembered that Triton, alone among large satellites in the Solar System, has retrograde motion.

MARS AT OPPOSITION

The 1984 opposition of Mars is not a very favourable one inasmuch as the minimum distance is much greater than is the case when Mars reaches opposition near the time of its perihelion. Northern observers are further handicapped by the fact that the planet is in Libra, well south of the celestial equator. However, small telescopes will show the main dark markings and the polar cap. The oppositions during the 1980s are:

	Day	*Max. apparent diameter, seconds of arc*	*Maximum magnitude*	*Constellation*
1980	Feb. 25	13.8	−1.0	Leo
1982	Mar. 31	14.7	−1.2	Virgo
1984	May 11	17.5	−1.8	Libra
1986	July 10	23.1	−2.4	Sagittarius
1988	Sept. 28	23.7	−2.6	Pisces

The next two oppositions will therefore be much more favourable – that of 1986 for Southern observers (Mars will be very low in the sky of Europe) and 1988 for observers everywhere, since the planet will be not far from the celestial equator.

THE FURTHEST PULSAR

Pulsars are rapidly-rotating neutron stars, sending out beams of pulsed radiation – which is how they were first identified; to date only three have been identified with optical objects (those in the Crab Nebula, in Vela, and the new exceptionally fast pulsar which seems to come into a rather different category). Pulsars are remnants of supernovæ, though it does not seem that all supernovæ must produce pulsars.

Until recently pulsar discoveries were confined to our Galaxy. However, in 1982 Dr John Ables, using the Parkes radio telescope in Australia, identified a pulsar in the Large Cloud of Magellan. This is an important breakthrough, and confirms that pulsars are to be found in outer systems – though in fact this had been assumed in any case; there was no logical reason to assume that they were confined to our Galaxy. Eventually, pulsars will no doubt be identified in the Andromeda Spiral and other systems millions of light-years away.

June

Full Moon: 13 June   *New Moon:* 29 June

Solstice: 21 June

Mercury is at superior conjunction on 23 June and is too close to the Sun for observers in the Northern Hemisphere. However, for those near to and south of the equator Mercury may be seen low above the east-north-east horizon in the mornings, about half-an-hour before sunrise, for the first ten days of the month.

Venus is at superior conjunction on 15 June and thus too close to the Sun for observation. Indeed for a short while on that date it could safely be said that no Earth-based instrument could detect it since it is actually occulted by the Sun.

Mars, its magnitude fading from −1.5 to −0.9 during the month, is still in Libra. It reaches its second stationary point on 20 June and then resumes its direct (eastward) motion. Mars is near Saturn and the minimum separation (4°.3) occurs on 11 June.

Jupiter is now at its brightest, magnitude −2.2, as it comes to opposition on 29 June and therefore available for observation throughout the night. It is a disappointing opposition for Northern Hemisphere observers because of its low altitude even when on the meridian (for example, the maximum altitude as seen from southern England is only 16°, the same as the midwinter Sun). When closest to the Earth its distance is 629 million kilometres. The path of Jupiter among the stars during its 1984 apparition is shown in Figure 6.

Saturn is visible in the evenings in Libra, magnitude +0.6. Although opposition occurred in May, its southern declination

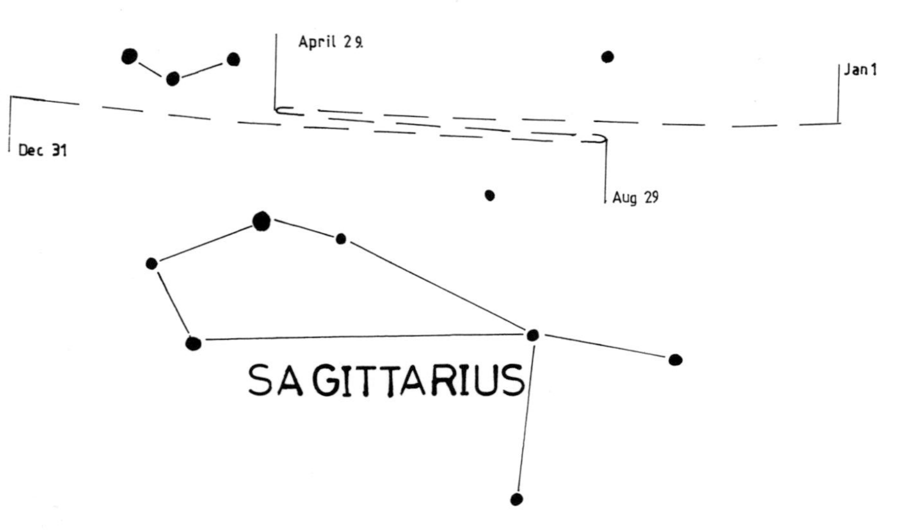

Figure 6. ***Path of Jupiter.***

means that, for observers in northern temperate latitudes, it is no longer visible after midnight by the end of the month.

Uranus is in the south-western part of Ophiuchus and is at opposition on 1 June, when it will be 2,693 million kilometres from the Earth. The magnitude of Uranus is +5.8 and therefore only just visible to the naked eye under good observing conditions. In a small telescope it exhibits a slightly greenish disk with a diameter of 4 arc seconds.

Neptune is at opposition on 21 June, in Sagittarius. It is not visible to the naked eye since its magnitude is +7.7. At opposition its distance from the Earth is 4,375 million kilometres.

URANUS AND NEPTUNE

Both the outer giant planets come to opposition this month. Though they are widely regarded as twins, there are important differences between them. At the moment our information is very incomplete, and we must hope for good results from Voyager 2, which is scheduled to by-pass Uranus in 1986 and Neptune in 1989. No further probes to these two planets have been planned by the United States as yet, and the Russians have so far sent no vehicles out beyond Mars, so that Voyager 2 seems to represent our only real chance of obtaining data from close range in the foreseeable future.

Of the two, it seems definite that Uranus is slightly the larger, but Neptune is appreciably the more massive. One major difference is that Uranus seems to have an internal heat-source, while Neptune does not. Also, the atmosphere of Uranus is clear to great depths, while that of Neptune contains aerosols.

Uranus has been found to have an extensive system of dark, narrow rings, not in the least like the magnificent ring-system of Saturn. Searches have been made for a Neptunian ring, but so far with negative results. It must be remembered that Neptune is unique in being attended by a very large satellite (Triton) with retrograde motion. This does not actually preclude the existence of a ring system, but it does make it considerably less likely.

The satellite systems, too, are different. Five Uranian satellites are known; all are smaller than our Moon, but four (Ariel, Umbriel, Titania, and Oberon) are of what may be termed 'medium' size, comparable with some of Saturn's satellites such as

Iapetus and Dione. Of Neptune's two confirmed satellites (a third has been suspected) Triton is certainly larger than our Moon, while Nereid has a highly eccentric orbit and is small – no more than 500 km in diameter.

Telescopes will show little upon the greenish disk of Uranus or the bluish face of Neptune. There is, too, the remarkable axial tilt of Uranus – not shared by Neptune – which amounts to 98 degrees, so that the rotation is technically retrograde. During the mid-1980s there will be polar presentation, so that the orbits of the satellites will appear virtually circular since all five revolve almost exactly in the equatorial plane.

EPSILON AURIGÆ

The prolonged eclipse finally ends on 25 June, so that for the next 27 years it may be assumed that the magnitude of the star will be constant at 2.99 – rather brighter than Eta Aurigæ (3.17).

MESSIER 13

This is a good time to look at the great globular cluster M.13, in Hercules. Its declination is +36°33′, so that it is best placed for northern observers, but it is also easy to see from countries such as Australia and South Africa.

Apart from Omega Centauri and 47 Tucanæ, both too far south to be seen from Europe, M.13 is the finest globular cluster in the sky. It was discovered by Edmond Halley in 1714, and is just visible with the naked eye, between the stars ζ and η Herculis, though it is not easy to find without optical aid. Binoculars show it well. Its distance is of the order of 21,000 light-years, and from it our Sun would appear as a dim star of magnitude +19. The overall diameter of M.13 seems to be about 160 light-years; the central region has a diameter of about 100 light-years (roughly a million cubic light-years in volume). The total number of stars in the cluster is uncertain, but may be as high as one million. Only a few variable stars have been discovered; the present count is four RR Lyræ variables, three Cepheids, and several Mira-type stars.

Spectroscopic measurements indicate that at present M.13 is approaching us at about 150 miles per second. However, this will not continue indefinitely; the movement is a combination of the Sun's motion in space, the rotation of the Galaxy, and the rotation of M.13 itself around the galactic centre. Rough estimates indicate that the cluster takes about 200,000,000 years to complete one circuit.

Small telescopes will resolve the outer parts of M.13, and the spectacle is glorious when seen through a larger instrument. If our Sun lay near the centre of the cluster the night sky would be brilliant; many stars would be bright enough to cast shadows, and there would be no darkness at all.

Nine degrees north-east of M.13 is another globular cluster, M.92, which is visible in binoculars about 6 degrees north of the star π Herculis. It is further away than M.13, but is intrinsically brighter, and as seen from Earth there is not much difference between the two.

July

Full Moon: 13 July *New Moon:* 28 July

Earth is at aphelion (furthest from the Sun) on 3 July at a distance of 152 million kilometres.

Mercury is now moving away from the Sun and reaches its greatest eastern elongation (27°) on 31 July. Although unsuitably placed for observation by those in northern temperate latitudes it is visible to observers further south for the last three weeks of the month, in the west-north-western sky, in the evenings, fading gradually during this period from magnitude −0.5 to +0.5.

Venus is unsuitably placed for observation being so close to the Sun, though at the very end of the month observers near the equator may glimpse it in the evenings for a very short while after sunset.

Mars, magnitude −0.6, is now well past opposition and no longer available for observation after midnight. Mars is still in Libra.

Jupiter, magnitude −2.2 is a conspicuous evening object and, indeed, is visible for the greater part of the night.

Saturn continues to be visible as an evening object in Libra, magnitude +0.7. The rings of Saturn continue to open wider and wider, the northern side being visible from the Earth and illuminated by the Sun.

BESSEL AND 61 CYGNI

This month marks the centenary of one of the most famous of all German astronomers, Friedrich Wilhelm Bessel, who was born at Minden on 22 July 1784. He carried out many important investigations, but is best remembered as being the first to measure the distance of a star.

Bessel began his career as a clerk in a commercial house in Bremen, and studied astronomy in his spare time. He computed the orbit of Halley's Comet, which had last returned to perihelion in 1759, and this work brought him to the notice of Dr Olbers, a leading German amateur. Olbers introduced him to Johann Schröter, who had his private observatory at Lilienthal, near Bremen, and Bessel became Schröter's assistant. In 1810 he moved to Königsberg as Professor of Astronomy and director of the Observatory there. His first important task was to carry out the deduction of Bradley's observations of star positions, and he also determined the positions of 75,000 stars.

Bessel was interested in star-distances, and decided to make a determined effort to measure a stellar parallax. The star he selected was 61 Cygni, which appeared to be a likely candidate because it was a wide binary and also had a large proper motion. In December 1838 he announced his result: the distance of 61 Cygni is now known to be just over 11 light-years. At about the same time the parallaxes of two more stars were announced; those of α Centauri (by Thomas Henderson, from his observations made at the Cape) and Vega (by Struve, from Dorpat). Though Henderson had made his actual observations some years earlier, there can be no doubt that the honour of priority must go to Bessel.

He then turned his attention to the irregular proper motions of Sirius and Procyon, and predicted that each star would be found to have a faint companion. In this he was correct, though the companions were not found until after Bessel's death. Bessel also became interested in the movements of Uranus, which he believed to be due to the action of a more distant planet. Together with his pupil, Flemming, he began work; but Flemming's early death, together with Bessel's ill-health, prevented much progress from being made. Bessel died at Königsberg on 17 March 1846, the year in which Neptune was located from the calculations made in France by Le Verrier.

IOTA^1 SCORPII

Scorpio or Scorpius, the Scorpion, is well placed this month. It lies well to the south of the celestial equator, and though its leading star, Antares, is easily observable from Europe the Scorpion's 'sting' barely rises over England and is out of view from the northern parts of the British Isles. The 'sting' contains one star, Scorpii or Shaula, which is of magnitude 1.63, and therefore only just too faint to be included among the stars conventionally regarded as being of the first magnitude.

Close to the 'sting' is a third-magnitude star, ι (Iota^1) Scorpii, which has a spectrum of type F2. It looks obscure enough, but it is in fact one of the most powerful of all the naked-eye stars. According to recent estimates, its distance is over 5,000 light-years, and its total luminosity must approach 200,000 times that of the Sun, so that intrinsically it is far more powerful than well-known 'cosmic searchlights' such as Rigel and Deneb. Were it as close as α Centauri, which shines as the third brightest star in the sky, ι Scorpii would cast shadows. Appearances can be deceptive!

August

Full Moon: 11 August *New Moon:* 26 August

Mercury continues to be visible in the western sky in the evenings for the first three weeks of the month, though not to observers in northern temperate latitudes. Mercury is fading during this period from magnitude +0.6 to +2.0, as it moves rapidly towards the Sun, passing through inferior conjunction on 28 August.

Venus is now moving outwards from the Sun and is visible as an evening object, magnitude −3.3. It is still a difficult object for observers in northern temperate latitudes who will only see it for a very short while after sunset, low above the western horizon.

Mars, magnitude around −0.1, continues to be visible to observers in the south-western sky in the evenings. Mars, which has been in the constellation of Libra since January, is moving eastwards more rapidly and during the month passes into the northern part of Scorpius.

Jupiter is visible in the evenings (magnitude −2.1) in the constellation of Sagittarius. It reaches its second stationary point at the end of August and then resumes its eastward motion. It is still well placed for observation for those in the Southern Hemisphere but to observers in northern temperate latitudes it will no longer be visible after 22^h by the end of August.

Saturn, magnitude +0.8, is visible in the south-western skies in the evenings.

THE PERSEIDS

Of all meteor showers, the Perseids are much the most reliable, and never fail to produce a good display. Unfortunately, in 1984 the Moon is full just at the time of maximum, and so the best part of the spectacle will be lost; however, many Perseids are bright, and the shower will be much in evidence all through the first fortnight of the month.

LUNAR PROBES

The first attempted American lunar probe, Able 1, was launched 26 years ago this month – on 17 August 1958. It was not a success; the lower stage of the launcher exploded after 77 seconds, at an altitude of 20 kilometres. There followed a further succession of failures, and success came only in July 1964, when Ranger 7 impacted the Moon after having sent back 4,308 close-range pictures of the Mare Nubium area.

August 1966 saw the launch of the first Orbiter probe, which was sent up on the 10th of the month. It was put into a closed path round the Moon, and sent back thousands of high-quality pictures. Four more Orbiters followed, all of which were successful and which provided excellent maps of virtually the whole of the lunar surface. Without these results, the Apollo missions would not have been possible. Orbiter 5, last of the series, was launched on 1 August 1967. Another August launching, this time a Russian one, was that of Luna 11 on 24 August 1966. This was the second Soviet lunar satellite (the first was Luna 10 of the previous March). On 9 August 1976 the Russians launched Luna 24, which landed on 18 August in the Mare Crisium, drilled down to 2 metres and obtained samples, lifting off again on the 19th and landing back on Earth on the 22nd. At present both the Russian and American lunar probe programmes are in abeyance, though it is reasonable to assume that further missions will be sent there in the foreseeable future.

THE SOUTHERN CROWN

There are two ‘crowns’ in the constellations, both included in Ptolemy’s original list of 48. Corona Borealis, the Northern Crown, is conspicuous both because of its shape and because its leading star, Alphekka, is of the second magnitude. The Southern Crown, Corona Australis, is much less prominent. It adjoins Sagittarius, and but for its fairly distinctive shape would certainly have been included in that constellation.

There are no stars above the fourth magnitude; the magnitudes of the two leaders, α and β are in each case 4.11, and there are only three more stars (γ, δ, and θ) above magnitude 4.7. γ is a close binary, with components which are rather unequal; the separation is only about 2 seconds of arc, and the period is of the order of 90 years.

The most interesting object is the variable nebula NGC 6729, whose nucleus contains the irregular variable R Coronæ Australis. Variations in the nebula generally follow changes in the star. The overall impression is similar to that of the more famous nebula associated with R Monocerotis (Hubble's variable nebula).

Corona Australis lies in the southern part of Sagittarius, close to α Sagittarii (Rukbat). It is interesting to note that in Sagittarius, the stars lettered α and β are both comparatively faint – below the fourth magnitude – while ε Sagittarii (Kaus Australis) is above the second magnitude, and σ(Nunki) only just below. This is one case where the star lettered α is by no means the leader of its constellation; another example is Corvus.

September

Full Moon: 10 September   *New Moon:* 25 September

Equinox: 22 September

Mercury is now emerging from the morning twilight and becomes visible to Northern Hemisphere observers after the first week of the month, remaining visible for approximately a fortnight. In fact, for these observers, this is the most suitable morning apparition of 1984, despite its relatively short duration. Figure 7 shows changes in azimuth (true bearing from the north through east, south, and west) and altitude of Mercury on successive mornings when the Sun is 6° below the horizon in latitude N.52°. At this time of year this condition, known as the beginning of morning civil twilight, occurs about 35 minutes before sunrise. The changes in brightness of Mercury are indicated approximately by the sizes of the circles which mark its position at five-day intervals. It will be noticed that Mercury is brightest after it reaches greatest western elongation.

Venus continues to be visible in the evenings shortly after sunset low above the western horizon. Northern Hemisphere observers will still only experience a relatively short period of observation after sunset since its motion away from the Sun is counteracted by its rapid southward movement in declination.

Mars, magnitude +0.3 remains visible in the south-western sky in the evenings, passing north of Antares at the beginning of September (see Figure 1 in February notes). At the end of September Mars passes into Sagittarius.

Jupiter is still a conspicuously bright object, magnitude −1.9, and visible in the south-western sky in the evenings.

Saturn has now faded to a magnitude of +0.9 and for observers in northern temperate latitudes, is only visible for a short while above the south-western horizon in the early evenings: by the end of the month it is only visible to observers in equatorial and southern latitudes.

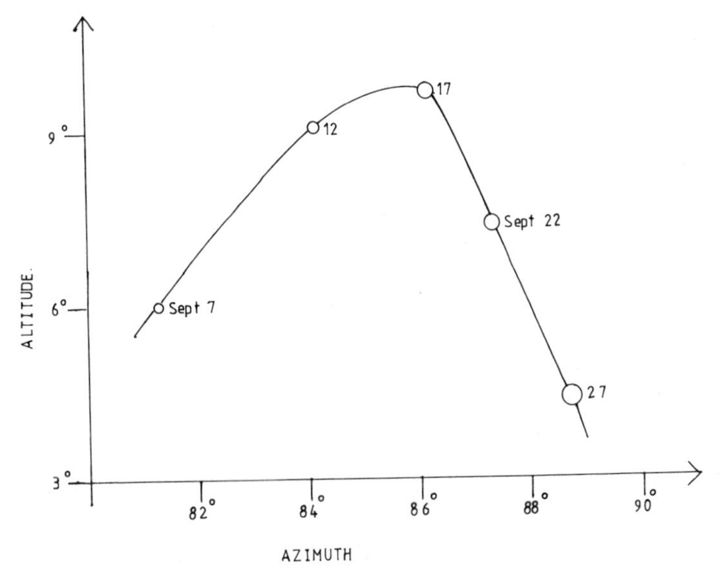

Figure 7. Mercury for N.52°.

THE BRIGHTNESS OF MERCURY

Mercury is much the least conspicuous of the planets known in ancient times. Yet it has been known since very early days; the oldest observation which has come down to us relates to 15 November 265 B.C. when, according to Ptolemy, the planet was one lunar diameter away from a line joining the stars δ and β Scorpii. Ptolemy also commented upon the yellowish colour of Mercury. Other observers have described it as pink, though most people will certainly term it white. Telescopes will show little upon its surface, and virtually all our knowledge about the features has been drawn from the probe Mariner 10, which made three active passes of the planet – in March and September 1974, and March 1975. There is an excessively tenuous atmosphere and a definite magnetic field. The phases were first detected by Hevelius in the seventeenth century.

Because Mercury is so elusive, there are many people who have never seen it (though the story that Copernicus never managed to observe it is probably untrue). The only reason for its lack of prominence is its closeness to the Sun. In fact Mercury is quite bright; its maximum magnitude is −1.9, which is brighter than any of the stars – even Sirius. Unfortunately it can never be seen against a dark background. Yet once it has been found with the naked eye, one tends to wonder how it can have been overlooked! It can often be found by sweeping the sky with binoculars, *but this should on no account be attempted unless the Sun is completely below the horizon.*

MARS AND ITS RIVAL

This is a good time to compare Mars with the bright red giant star Antares, whose very name means 'the Rival of Ares' (Ares being the Greek equivalent of the war-god Mars). At present Mars, at magnitude 0.3, is the brighter of the two; the magnitude of Antares is 0.96 (though, like so many red giants, it is very slightly variable). However, the colours are much the same. Once again appearances are deceptive. Mars is one of the smaller members of the Sun's family of planets; Antares is large enough to hold the entire orbit of the Earth round the Sun. Its distance is over 300 light-years, and its luminosity is some 7,500 times that of the Sun.

ALPHA CAPRICORNI AND OTHER NAKED-EYE DOUBLES

Capricornus, the Sea-goat, is well placed this month. It is one of the less prominent Zodiacal constellations; its brightest star, δ, is only of the third magnitude, and there is no really distinctive shape. However, α Capricorni, also known by its proper name of Al Giedi, is interesting because it is a naked-eye double. α^1 is of magnitude 4.2, while α^2 is rather brighter at magnitude 3.6. The separation is 376 seconds of arc, and there is no difficulty in splitting the pair. They are not genuinely associated: α^1 is 1,600 light-years from us; α^2 only 117. Both are again double. α^1 has a 9th-magnitude optical companion at a distance of 45″.5 (position angle 221°); α^2 has a companion of magnitude 10.6 at 7″.1, position angle 158°. To complete the picture, the fainter member of the α^2 pair is a close binary with a separation of 1″.2.

Naked-eye doubles are rather rare. The classic case is that of Mizar (ζ Ursæ Majoris), whose 5th-magnitude companion, Alcor, is an easy object. In the Hyades we have θ Tauri, a very wide pair, and in Lyra there is the celebrated ε Lyræ, which can be separated on a good night by anyone with normal eyesight; the separation is 208″, and each star is again double. There are various doubles which are separable with binoculars; one is ν Draconis in the Dragon's head, and another – probably the most beautiful double in the sky – is β Cygni or Albireo, with its golden-yellow 3rd-magnitude primary and its blue 5th-magnitude companion. The separation for Albireo is approximately 35 seconds of arc.

In the far south of the sky, well placed near the zenith this month for Australian and South African observers, is the constellation of Grus, the Crane. Both δ and μ Gruis look like naked-eye doubles, though the separations are so great that they are not usually regarded as true pairs.

October

Full Moon: 9 October   *New Moon:* 24 October

Summer Time in Great Britain and Northern Ireland ends on 28 October.

Mercury passes through superior conjunction on 10 October. As a result it is virtually impossible to observe the planet with the naked eye at all during the month. The only exception to this statement is that on the last two evenings of the month keen sighted observers in the tropics and in the Southern Hemisphere might glimpse the planet very low above the west-south-western horizon at the end of evening civil twilight.

Venus, magnitude −3.4, continues to be visible in the evenings, low in the south-western sky after sunset.

Mars is still over 70° from the Sun and still visible in the south-western skies in the evening. For observers in northern temperate latitudes it will only be seen at a low altitude above the south-western horizon, since Mars reaches a declination of almost −26° during the month. Mars passes 2°S. of Jupiter late on 13 October, though for observers in the British Isles the time of closest approach occurs after they have set. A pleasing spectacle can be witnessed on the evening of 29 October as the crescent Moon passes south of Mars, having previously passed south of Jupiter earlier on the same day.

Jupiter, magnitude −1.7, is an evening object in the south-western sky in the early evenings.

Saturn continues to move in towards the Sun (apparently!) and even for Southern Hemisphere observers is lost in twilight before the end of the month.

THE LIMB OF THE MOON

More than a quarter of a century ago, in October 1959, the Russian unmanned probe Lunik (or Luna) 3 went round the Moon, sending back the first pictures of the hidden side. By the standards of today the Lunik pictures look very blurred and difficult to interpret, though in 1959 they represented a tremendous triumph. Several features were discovered, including the grey area which the Russians named the Mare Moscoviense (Moscow Sea) and the great dark-floored crater now known as Tsiolkovskii in honour of Konstantin Eduardovich Tsiolkovskii, the Russian pioneer of theoretical rocketry who was writing sound sense about space-flight well over seventy years ago. Other features were, however, misinterpreted, and the mountain chain which the Russians named the Soviet Mountains does not exist at all; it is nothing more nor less than a bright ray, and the name has been deleted.

Because the Moon's orbit round the Earth (or, to be more accurate, the barycentre or centre of gravity of the Earth–Moon system) is not circular, the position in orbit and the amount of axial rotation become periodically 'out of step', so that we can see a little way around alternate east and west limbs. All in all, Earth-based observers can examine 59 per cent of the Moon's surface, though of course, never more than 50 per cent at any one moment. Only 41 per cent remains permanently out of view, despite the various 'tilts' or librations.

However, the regions close to the Moon's limb as seen from Earth are so foreshortened that they are difficult to map – as pre-Space Age lunar observers found only too well. All formations away from the centre of the apparent disk show a degree of foreshortening. For instance, the dark-floored, 60-mile crater Plato, near the foothills of the Lunar Alps, appears oval, though in fact it is almost perfectly circular – and was shown as such by the Orbiter and Apollo pictures of it. The Mare Crisium or Sea of Crises, near the Moon's limb (to the upper right as seen with the naked eye from the Northern Hemisphere of the Earth) appears to be elongated in a north–south direction; actually it is slightly elongated in an east–west direction. Close to the limb it is hard to tell a crater from a ridge. There are also some 'seas' which are very foreshortened. One of these is the Mare Orientale or Eastern Sea, which was first noted and named by H. P. Wilkins and the Editor of this *Yearbook* more than thirty years ago. There was then no way of telling that the Mare Orientale is an important and very large ringed feature, extending well on to the permanently averted part of the Moon.

Before the age of space research, lunar cartographers were very busy in doing their best to map the foreshortened areas. Looking back, it is permissible to say that they managed fairly well; but of course this sort of work is now obsolete, and the probe pictures have provided us with detailed maps of practically the whole of the lunar surface.

DIPHDA AND SUSPECTED VARIABLES

The brightest star in the constellation of Cetus (the Whale) is β, or Diphda, with a magnitude of 2.04. It has been suspected of variability, but the fluctuations have not been confirmed, and their reality is doubtful. It is interesting to look back at a catalogue of suspected variables published in a famous book, *Handbook of Astronomy*, by G. F. Chambers in 1890. Some of the entries are:

Star	*Range*	*Authority*
γ Pegasi	2½–3	Schwabe, Period 27½d?
ν Fornacis	5–6	Gould
γ Eridani	2½–3½	Secchi
σ Canis Majoris	4½–5	Gould
β Volantis	4–5	Gould
η Crateris	4½–6½	Houzeau. 1875
δ Ursæ Majoris	2½–4	Pigott
ε Corvi	3–4	Gould
η Virginis	3–4	Gould
γ Corvi	2¾–3½	Gould
η Ursæ Majoris	± 2	Espin and others
ν Boötis	4–4½	Schmidt
θ Apodis	5½–6½	Gould
β Ursæ Minoris	2¼–2¾	J. Herschel; Espin. Period 10d?
μ Draconis	4–5	S. J. Johnson
ι Apodis	5–6	Gould
γ Sagittarii	3–3½	Gould. Period l^{D}ng?
β Cygni	3–4	Klein, Period: years?
μ Aquilæ	4–5	Gould
ε Draconis	3¾–4¾	Gould
ρ Pavonis	4½–5¼	Gould
ν Pavonis	5–6	Gould
γ Indi	6–6½	Gould

β Cephei	3–3½	Schwabe. Period 383d?
ε Pegasi	2–2½	Schwabe. Period 25½d?
ζ Piscis Aust.	5–6½	Schmidt. Period long?
η Pegasi	3–3¼	Christie
ι Cephei	3¾–4½	Schmidt. Period 369d?
ι Andromedæ	4½–5	Gore

Variations prominent enough to be noticed with the naked eye have not been confirmed for any of these stars, and probably do not occur, but it is an interesting list none the less. Incidentally, one star – α Cassiopeiæ or Shedir – long regarded as variable was listed by Kukarkin as 'constant', though it does seem that there are slight fluctuations between magnitude 2.1 and 2.4.

November

Full Moon: 8 November *New Moon:* 22 November

Mercury attains its greatest eastern elongation (22°) on 25 November and thus is visible in the evenings to observers in the tropics and the Southern Hemisphere for the whole of the month. This evening apparition, which continues through to early December is the most suitable one of the year for observers in southern latitudes. Figure 8 shows the changes in azimuth and altitude of Mercury on successive evenings when the Sun is 6° below the horizon in latitude S.35°. At this time of year, this condition, known as the end of evening civil twilight, occurs about 30 minutes after sunset. The changes in brightness are roughly indicated by the sizes of the circles which mark Mercury's position at five-day intervals. It will be seen that Mercury is brightest before it reaches eastern elongation. During November Mercury fades in brightness from magnitude −0.4 to +0.3.

Venus, magnitude −3.5, continues to be visible above the south-western horizon after sunset. Venus passes 2°S. of Jupiter on 24 November.

Mars, magnitude +0.8, is still an evening object though not well placed for observers in northern temperate latitudes because of its low altitude above the south-western horizon.

Jupiter remains visible in the south-western sky in the early part of the evening.

Saturn is in conjunction with the Sun on 11 November and thus unavailable for observation.

A total eclipse of the Sun will occur on 22–23 November, see page 121 for details.

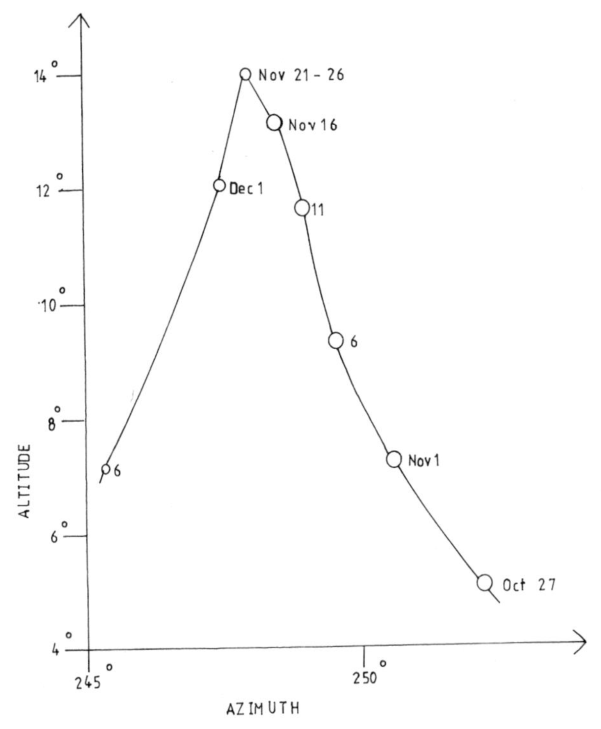

Figure 8. Mercury for S.35°.

CERES

The largest and first-discovered of the minor planets, Ceres, comes to opposition this month (10 November). As its magnitude is only 6.9 it is too faint to be seen with the naked eye, but binoculars will show it, though of course it looks exactly like a star. It was discovered by Piazzi on 1 January 1801, the first day of the new century. (Looking ahead, it is worth noting that the first day of the coming century will be 1 January 2001 – not 1 January 2000, as so many people imagine.)

Ceres has a diameter estimated at 623 miles. It is much the largest of the asteroids, and accounts for more than half the total mass of the swarm. No other asteroid is as much as 400 miles across; so far as we know, only Pallas, Vesta, and Hygeia exceed 230 miles. Ceres has a rotation period of 9 hours. Its revolution period is 4.06 years.

ANDERS JOHN LEXELL AND THE COMET OF 1770

Lexell, born at Turku (Abo) in 1740, is one of the best-remembered Finnish astronomers. He was Professor of Mathematics at Turku University from 1775 to 1780, though he actually stayed in St Petersburg, where he had been appointed to the Academy of Science in 1769. He took part in the observations of the 1769 transit of Venus, giving a value for the solar parallax of 8″.63.

Lexell computed an orbit for the comet of 1769, and then turned his attention to the comet of 1770, which had been discovered by Charles Messier on 14 June. It was visible with the naked eye, and its apparent diameter was given as five times greater than that of the full Moon – not because of exceptional size, but because it was exceptionally near; it passed within 750,000 miles of the Earth, which still remains a record for comets. Lexell calculated a period of 5½ years. Before 1767 the period had been much longer, but then a close encounter with Jupiter altered the orbit which brought it close to the Earth.

Presumably the next return would have taken place in 1775–6, but it was not observed, and unfortunately Lexell did not publish his results until 1778, by which time the situation had changed again. Before the next return there was a second encounter with Jupiter – in 1779 – and the new orbit meant that the comet came nowhere near the Earth. It has never been seen again, and there seems no hope of its being recovered now. We remember it as Lexell's Comet.

In 1781 William Herschel reported to the Royal Society that he had discovered a comet in the constellation of Gemini. Orbits were computed independently by several mathematicians, including de Saron, in France, and Lexell. Both found that the object was much further away than Saturn, and was in fact a planet – the world we now know as Uranus.

Lexell carried out several more important researches, and was active until shortly before his death on 30 November 1784, at St Petersburg – two hundred years ago.

ACHERNAR AND THE SOUTH POLE

Achernar, or α Eridani, is much too far south to be seen from anywhere in Europe, but as seen from Australia or South Africa it is almost overhead during November evenings. It is the ninth brightest star in the sky, with a magnitude of 0.46; its spectral type is B5, so that it is pure white; it lies at a distance of 85 light-years, and is almost 800 times as luminous as the Sun.

The south celestial pole is not marked by any bright star, and lies in a rather blank area in the constellation of Octans (the Octant). It so happens that Achernar and the Southern Cross lie on opposite sides of the pole, and at about the same distance from it, which is probably the best way to locate the polar area. During November evenings, when Achernar is so high up, the Cross is skirting the southern horizon.

December

Full Moon: 8 December *New Moon:* 22 December

Solstice: 21 December

Mercury continues to be visible in the evenings for the first week of December, though not to observers in northern temperate latitudes. During this week its magnitude fades from +0.3 to +1.0. It passes through inferior conjunction on 14 December and then moves rapidly west of the Sun so that it becomes visible as a morning object for the last ten days of the month, its magnitude increasing from +1.0 to 0.0 during this period. Since Mercury is almost stationary about 10° from Antares (magnitude +1.0), during this period observers will have a good opportunity of watching the increasing brightness of Mercury relative to Antares.

Venus is still visible as a magnificent object in the evening skies. As it moves northwards in declination Northern Hemisphere observers obtain a longer period of observation and in fact Venus is visible for over three hours after sunset by the end of the year.

Mars is now moving rapidly eastwards in Capricornus and then into Aquarius. It is no longer the conspicuous object it was at opposition, its magnitude having faded to +1.0. It is still visible in the evenings in the south-western sky.

Jupiter is approaching the end of its 1984 apparition and only visible for a short while in the early evenings low above the south-western horizon. Because of its southern declination observers in northern temperate latitudes will only be able to see it for about the first week or ten days of the month.

Saturn, magnitude +0.8, becomes visible as a morning object early in the month. It may be detected low above the south-eastern horizon before the sky gets too bright for observation. Saturn is still in Libra.

SU TAURI

The variable star SU Tauri, not far from the boundary between Taurus and Orion, is never conspicuous. At its brightest it may attain magnitude 9½, but this is its maximum, and more generally the magnitude is around 9.7. It is a member of the rare class known as R Coronæ stars, since R Coronæ Borealis is the brightest and best-known of the type.

These variables remain at maximum for protracted periods, with only slight fluctuations, but then suffer sudden, quite unpredictable fades. At minimum SU Tauri may descend to magnitude 16, so that it is then beyond the range of all but powerful instruments. It is known that R Coronæ stars are deficient in hydrogen but very rich in carbon, and it is thought that the fades are due to the accumulation of carbon particles in the stars' atmospheres. The only five examples with maxima brighter than magnitude 10 are:

Star	*Max.*	*Min.*	*Spectrum*
S Apodis	9.6	15.2	R
UW Centauri	9.6	16.0	K
R Coronæ Borealis	5.8	15	Gp
RY Sagittarii	6.0	14.0	Gp
SU Tauri	9.5	16	G
RS Telescopii	9.3	14.6	R

Of these, only SU Tauri and R Coronæ itself are well placed for Northern Hemisphere observers. Amateur astronomers can make useful observations of them, and have often been the first to detect signs of fading in them.

HALLEY'S COMET

During 1984 Halley's Comet will brighten steadily as it approaches the Sun, though it is not likely to become a naked-eye object before November 1985 – and even then it will not be spectacular, since this is, unfortunately, a very poor return. During 1984 the comet will remain in the Gemini area, moving into the northern part of Orion by the end of December, but it will still be beyond the range of most amateur instruments. A detailed article about the comet will be given in the 1985 *Yearbook*. Perihelion is due on 9 February 1986.

THE SHORTEST-PERIOD COMET

Encke's Comet, which has returned this year, has a period of 3.3 years, and is the only periodical comet which can be followed throughout its orbit (apart from the exceptional comets with near-circular orbits, moving beyond Jupiter). However, in 1949 it was thought that the record had been broken. Comet Wilson-Harrington, discovered in that year, was calculated to have a period of only 2.3 years; but it has never been seen again, and must be regarded as permanently lost. So too is Helfenzrieder's Comet, seen for the first and only time in 1766; the period was calculated as 4.5 years. Of known comets, the nearest rival to Encke's is Grigg-Skjellerup, with a period of 5.1 years. It has now been seen at more than a dozen returns, and its orbit is so well known that there is no danger of its being lost.

NOAH'S DOVE

Columba Noachi, Noah's Dove, was one of the original 48 constellations listed by Ptolemy; it is known today simply as Columba. It is easy to find, south of Orion, though it is always very low as seen from Britain, and one of its brighter stars (η) does not rise at all, since its declination is almost 43 degrees south. The two brightest stars are α or Phakt (magnitude 2.64) and β or Wazn (3.12). There are only ten stars in the constellation brighter than the fifth magnitude.

μ Columbæ, of magnitude 5.2, is of interest because it is one of three early-type stars (spectral type 09.5) moving out at high velocities from the nebulous region of Orion; the other two are 53 Arietis and AE Aurigæ. All three have abnormally high space velocities. μ Columbæ has an annual proper motion of 0″.025; the true space velocity is around 74 miles per second. Its distance is over 2,700 light-years, and it is approximately 5,000 times as luminous as the Sun.

Eclipses in 1984

In 1984 there will be two eclipses of the Sun and none of the Moon.

(1) *An annular eclipse of the Sun on 30 May.* It is visible as a partial eclipse from the Hawaiian Islands, the eastern part of the Pacific Ocean, North America except the north-west, Central America, the West Indies, the extreme north-west of South America, the Arctic regions, the Atlantic Ocean, Greenland, Iceland, Europe (including the British Isles) and north-west Africa. The eclipse begins at $13^h\ 54^m$ and ends at $19^h\ 35^m$. The annular phase begins in the Pacific Ocean, crosses Mexico, the Gulf of Mexico, the south-eastern part of the United States of America, the Atlantic Ocean and Morocco and ends in Algeria.

From south-east England the partial phase of the eclipse begins at around $17^h\ 15^m$ and ends around $19^h\ 00^m$ while in Scotland it begins a few minutes earlier and has a duration of about 1½ hours. After this eclipse it will be another nine years before an eclipse of the Sun will be visible from anywhere in the U.K.

(2) *A total eclipse of the Sun on 22–23 November.* The path of totality begins in the Molucca Islands, crosses New Guinea and the Coral Sea, passes north of New Zealand and ends in the South Pacific Ocean. The partial phase is visible from the eastern part of Indonesia, the southern part of the Philippine Islands, New Guinea, Australia, Tasmania, Oceania, New Zealand, the South Pacific Ocean, part of Antarctica and the extreme south of South America. The eclipse begins at $22^d\ 20^h\ 13^m$ and ends at $23^d\ 01^h\ 33^m$; the total phase begins at $22^d\ 21^h\ 13^m$ and ends at $23^d\ 00^h\ 34^m$. The maximum duration of totality is $2^m\ 00^s$.

From south-western Australia the partial phase starts shortly after sunrise and lasts for just over an hour. For observers in south-east Australia the eclipse begins around 21^h and lasts for two hours while in New Zealand the eclipse begins about half-an-hour later and lasts for around 2½ hours.

Lunar Occultations in 1984

In the course of its journey round the sky each month, the Moon passes in front of all the stars in its path and the timing of these occultations is useful in fixing the position and motion of the Moon. The Moon's orbit is tilted at more than five degrees to the ecliptic, but it is not fixed in space. It twists steadily westwards at a rate of about twenty degrees a year, a complete revolution taking 18.6 years, during which time all the stars that lie within about six and a half degrees of the ecliptic will be occulted. The occultations of any one star continue month after month until the Moon's path has twisted away from the star but only a few of these occultations will be visible at any one place in hours of darkness.

There are a number of occultations in 1984, all the bright planets except Jupiter being occulted on at least one occasion. Unfortunately none of the events are visible from the British Isles. Saturn is occulted on several occasions as seen from one part or other of Australasia – on 20 March, 10 June, 31 August, and 27 September. Mars is occulted on 17 April, as seen from the west of Australia.

Only four first-magnitude stars are near enough to the ecliptic to be occulted by the Moon; these are Regulus, Aldebaran, Spica, and Antares. There is not a single occultation of a first-magnitude star visible in 1984.

Predictions of these occultations are made on a world-wide basis for all stars down to magnitude 7.5, and sometimes even fainter. Lunar occultations of radio sources are also of interest – remember the first quasar, 3C.273, was discovered as the result of an occultation.

Recently occultations of stars by planets (including minor planets) and satellites have aroused considerable attention.

The exact timing of such events gives valuable information about positions, sizes, orbits, atmospheres and sometimes of the presence of satellites. The discovery of the rings of Uranus in 1977 was the unexpected result of the observations made of a predicted occultation of a faint star by Uranus. The duration of an occultation by a satellite or minor planet is quite small (usually of the order of a minute or less). If observations are made from a number of stations it is possible to deduce the size of the planet.

The observations need to be made either photoelectrically or visually. The high accuracy of the method can readily be appreciated when one realizes that even a stop-watch timing accurate to $0^s.1$ is, on average, equivalent to an accuracy of about 1 kilometre in the chord measured across the minor planet.

Comets in 1984

The appearance of a bright comet is a rare event which can never be predicted in advance, because this class of object travels round the Sun in an enormous orbit with a period which may well be many thousands of years. There are therefore no records of the previous appearances of these bodies, and we are unable to follow their wanderings through space.

Comets of short period, on the other hand, return at regular intervals, and attract a good deal of attention from astronomers. Unfortunately they are all faint objects, and are recovered and followed by photographic methods using large telescopes. Most of these short-period comets travel in orbits of small inclination which reach out to the orbit of Jupiter, and it is this planet which is mainly responsible for the severe perturbations which many of these comets undergo. Unlike the planets, comets may be seen in any part of the sky, but since their distances from the Earth are similar to those of the planets their apparent movements in the sky are also somewhat similar, and some of them may be followed for long periods of time.

The following periodic comets are expected to return to perihelion in 1984:

Comet Taylor was discovered in 1915. Dr E. E. Barnard discovered the bifurcation of the comet's nucleus on 9 February 1916. Although it has a period of about 7 years it was not seen again until it was re-discovered in 1976.

Comet Crommelin was discovered by Pons in 1818 and has been observed at four apparitions since then. It has a period of 27 years. It was originally known as Comet Pons-Coggia-Winnecke-Forbes but was re-named in 1948 in honour of A. C. D. Crommelin whose work had established the identity of the three comets – Pons (1818), Coggia and Winnecke (1873), and Forbes (1928).

Comet Smirnova-Chernykh was discovered in 1975 and has a period of 8.5 years.

Comet Clark was discovered in 1973 and re-observed at its return in 1978. It has a period of 5.5 years.

Comet Faye was discovered in 1843 and has made 17 appearances. It has a period of 7.3 years.

Comet Tritton was discovered in 1978 and has a period of 6.4 years.

Comet Encke has the shortest known period (3.3 years) of any comet. It was first seen by Méchain in 1785 but it carries the name of the great mathematician Encke because he not only showed that the comets of 1785, 1795, 1805, and 1819 were returns of the same comet, but he also predicted its return in 1822 and this was duly observed. This was only the second prediction of the return of a periodic comet, the first being Halley's prediction of the return of the Comet Halley in 1759. Although the orbit of Encke's comet is very eccentric it is nowadays detected at each opposition by the world's largest telescopes. The return to perihelion in 1984 will be its 53rd.

Comet Wolf was discovered in 1884 and had made twelve returns to perihelion at its last appearance in 1975. The original orbit had a period of 6.8 years, but a close approach to Jupiter in 1922 enlarged the orbit so that the period became 8.4 years. This orbit is very well defined since it has been the subject of intensive study by the Polish astronomer Michael Kamienski (1879–1973), who was able to demonstrate the fact that the mean motion of the comet is slowly decreasing. This was one of the first examples known of the action of non-gravitational forces on the motion of a comet.

Comet Wild was discovered in 1978 and is making its first return. It has a period of 6.2 years.

Comet Wolf-Harrington was discovered in 1925 and has a period of 6.5 years.

Comet Neujmin was discovered in 1913 and has been seen at every return since then. It has a period of 18.2 years.

Comet Arend-Rigaux was discovered in 1951 and has a period of 6.8 years.

Comet Haneda-Campos was discovered in 1978 and is making its first return. It has a period of 6.3 years.

Minor Planets in 1984

Although many thousands of minor planets (asteroids) are known to exist, only 2,100 of these have well-determined orbits and are listed in the catalogues. Most of these orbits lie entirely between the orbits of Mars and Jupiter. All of these bodies are quite small, and even the largest, Ceres, is believed to be only about 1,000 kilometres in diameter. Thus, they are necessarily faint objects, and although a number of them are within the reach of a small telescope few of them ever reach any considerable brightness. The first four that were discovered are named Ceres, Pallas, Juno and Vesta. Actually the largest four minor planets are Ceres, Pallas, Vesta and Hygeia. Vesta can occasionally be seen with the naked eye and this is most likely to occur when an opposition occurs near June, since Vesta would then be at perihelion. In 1984 Ceres will be at opposition on 10 November with a magnitude of 6.9 and Pallas will be at opposition on 6 September with a magnitude of 8.8.

A vigorous campaign for observing the occultations of stars by the minor planets has produced improved values for the dimensions of some of them, as well as the suggestion that some of these planets may be accompanied by satellites. Many of these observations have been made photoelectrically. However, amateur observers have found renewed interest in the minor planets since it has been shown that their visual timings of an occultation of a star by a minor planet are accurate enough to lead to reliable determinations of diameter (see page 123). As a consequence many groups of observers all over the world are now organizing themselves for expeditions should the predicted track of such an occultation cross their country.

Meteors in 1984

Meteors ('shooting stars') may be seen on any clear moonless night, but on certain nights of the year their number increases noticeably. This occurs when the Earth chances to intersect a concentration of meteoric dust moving in an orbit around the Sun. If the dust is well spread out in space, the resulting shower of meteors may last for several days. The word 'shower' must not be misinterpreted – only on very rare occasions have the meteors been so numerous as to resemble snowflakes falling.

If the meteor tracks are marked on a star map and traced backwards, a number of them will be found to intersect in a point (or a small area of the sky) which marks the radiant of the shower. This gives the direction from which the meteors have come.

The following table gives some of the more easily observed showers with their radiants; interference by moonlight is shown by the letter M.

Limiting dates	*Shower*	*Maximum*	*R.A.*	*Dec.*	
Jan. 1–6	Quadrantids	Jan. 4	15^h28^m	$+50°$	
Mar. 14-18	Corona Australids	Mar. 16	16^h20^m	$-48°$	M
April 20–22	Lyrids	April 21	18^h08^m	$+32°$	M
May 1–8	Aquarids	May 5	22^h20^m	$+00°$	
June 17–26	Ophiuchids	June 19	17^h20^m	$-20°$	
July 15–Aug. 15	Delta Aquarids	July 27	22^h36^m	$-17°$	
July 15–Aug. 20	Pisces Australids	July 30	22^h40^m	$-30°$	
July 15–Aug. 25	Capricornids	Aug. 1	20^h36^m	$-10°$	
July 27–Aug. 17	Perseids	Aug. 12	3^h04^m	$+58°$	M
Oct. 15–25	Orionids	Oct. 20	6^h24^m	$+15°$	
Oct. 26–Nov. 16	Taurids	Nov. 7	3^h44^m	$+14°$	M
Nov. 15–19	Leonids	Nov. 17	10^h08^m	$+22°$	M
Dec. 9–14	Geminids	Dec. 14	7^h28^m	$+32°$	M
Dec. 17–24	Ursids	Dec. 22	14^h28^m	$+78°$	

M = moonlight interferes

Maxima of Long-Period Variable Stars, 1984

Stars of the Mira type have periods and amplitudes which are not completely regular, so that it is not possible to predict their maxima with complete accuracy. Moreover, a Mira star changes slowly, and may appear to remain at maximum for several consecutive nights. The following predictions for 1984 have been provided by D. R. B. Saw, Director of the Variable Star Section of the British Astronomical Association.

Star	*Desig.*	*Maximum*
R And	0018 + 38	Nov. 29
W And	0211 + 43	July 16
R Aql	1901 + 08	Jan. 10 Oct. 18
R Ari	0210 + 24	Apr. 15 Oct. 19
X Aur	0604 + 50	Feb. 6 July 19
R Boö	1432 + 27	Jun. 7
U Boö	1449 + 18	July 8
V Boö	1425 + 39	Apr. 21
V Cam	0549 + 74	Oct. 29
X Cam	0432 + 74	May 7 Sept. 27
S Cas	0112 + 72	Jan. 21
T Cas	0017 + 55	Sept. 24
T Cep	2108 + 68	Feb. 8
o Cet	0214 – 03	May 9
S CrB	1517 + 31	Dec. 19
V CrB	1546 + 39	Aug. 1
W CrB	1611 + 38	Jun. 6
R Cyg	1939 + 49	Jun. 20
S Cyg	2003 + 57	Jan. 5 Nov. 22
Chi Cyg	1946 + 32	May 25
R Dra	1632 + 66	Mar. 4 Nov. 3
T Dra	1754 + 58a	Dec. 10
RU Her	1606 + 25	Jun. 17
SS Her	1628 + 07	Feb. 18 Jun. 4 Sept. 19
R Hya	1324 – 22	Sept. 5
R Leo	0942 + 11	Jan. 25 Dec. 4
W Lyr	1811 + 36	May 27 Dec. 9
U Ori	0549 + 20a	Nov. 1
R Ser	1546 + 15	Jun. 21
R Tri	0231 + 33	Feb. 19 Nov. 11
T UMa	1231 + 60	July 6
S Vir	1327 – 06	Apr. 21

Some Events in 1985

ECLIPSES

In 1985 there will be four eclipses, two of the Sun and two of the Moon.

4 May: total eclipse of the Moon – Australasia, Asia (except the north and north-east), Africa, Europe, and extreme eastern South America.

19 May: partial eclipse of the Sun – north-eastern Asia, northern North America, north-west Europe.

28 October: total eclipse of the Moon – Alaska, Australasia, Asia, Africa, Europe.

12 November: total eclipse of the Sun – southern South America.

THE PLANETS

Mercury may be seen more easily from northern latitudes in the evenings about the time of greatest eastern elongation (17 March) and in the mornings around greatest western elongation (28 August). In the Southern Hemisphere the corresponding dates are around 3 January (mornings) and 8 November (evenings).

Venus is visible in the evenings at the beginning of the year. After inferior conjunction on 3 April it is visible in the mornings for the rest of the year.

Mars does not come to opposition in 1985.

Jupiter is at opposition on 4 August.

Saturn is at opposition on 15 May.

Uranus is at opposition on 6 June.

Neptune is at opposition on 23 June.

Pluto is at opposition on 23 April.

PART TWO

Article Section

The New Observatory on La Palma, Canary Islands

PAUL MURDIN and ALEC BOKSENBERG

At sunrise on 9 December 1982 the SS *Alraigo* tied up at the dockside of the city of Santa Cruz on the island of La Palma, most westerly of the Canary Islands. Within hours there was winched from its hold a cylindrical metal box, 3.2 metres diameter, 90 centimetres thick and weighing 6½ tons. On the red-brown sides, the box was labelled with the conventional wine glass symbol and the understatement GLASS – DO NOT DROP; it was also stencilled with the outline of a portly engineer peering skywards through a telescope standing on the letters 'RGO', a trademark and identifying symbol for equipment sent to La Palma from the Royal Greenwich Observatory. It was swung on to the low-level trailer parked among the bananas stacked on the dockside and began its journey through the city, and up towards the central peak of the island, the mountain known as Roque de los Muchachos, after the formation of pillars at the top, reminiscent of a friendly huddle of companions.

Thus arrived on La Palma the new 2.5-metre mirror of the Isaac Newton Telescope, first of three British telescopes to be operational in the international Observatorio del Roque de los Muchachos.

Next to be completed will be a 1-metre telescope, and in 1986 the 4.2-metre William Herschel Telescope, which then will be the fourth largest in the world. The Herschel Telescope, presently under manufacture, will be the largest telescope presently foreseen in an international observatory formed by Spain, the United Kingdom, Sweden, and Denmark, with the further participation of the Netherlands and Eire. Six telescopes are presently planned.

The Observatorio del Roque de los Muchachos is at a height of 2,400-metres on a site chosen for the clarity of its atmosphere and excellent *seeing*. Seeing in astronomy means the amount by which

Figure 1. The new 2.5 m mirror for the Isaac Newton Telescope is lifted from the hold of a ship, dockside La Palma in the Canary Islands. (Photo: Rutherford-Appleton Laboratory.)

light from distant point-like stars is blurred in its passage through the Earth's atmosphere. Good seeing means less blurring. Seeing occurs because the Earth's atmosphere is made turbulent by movement of the air with bubbles of hot air rising into colder layers. Just as the distant view is blurred when seen through a misty windscreen covered with droplets of water, so stars are blurred by the atmospheric turbulence. Warmed by the Sun, the surface of the sea drives the air upwards by convection, a process which forms cloud and would, even if clear, give bad seeing. But La Palma is down wind of the cold Canary current, and air passing over the cold current is stabilized and flows uniformly around the mountain. Thus seeing at the Roque de los Muchachos is excellent, and for nearly half the time star images are blurred by 1 arc sec or less.

For comparison, seeing at a sea-level site (e.g. everywhere in Britain) is typically 5 arc sec or more (1 arc sec is the angle subtended by a two-centimetre diameter coin at a distance of 4 kilometres).

The importance of good seeing for astronomy is that faint star images are detected against a background of light from the night sky. The worse the seeing, the more the starlight is spread out, and the more background contamination there is. Good seeing means detecting fainter and more distant stars. Seeing further than before makes parts of the universe accessible which were otherwise closed to our study.

In designing and building the United Kingdom telescopes on La Palma, astronomers and engineers at the Royal Greenwich Observatory, in consultation with Freeman Fox and Partners and with the telescope builders, Sir Howard Grubb Parsons Ltd, are meeting the challenge of producing telescopes which will exploit the ideal conditions on La Palma. They work on behalf of

Figure 2. In what is rumoured to be a self-portrait by RGO engineer Reg Stokoe, a portly man looks at the sky over the island of La Palma. He looks through a telescope founded on the island by the RGO. This logo identifying equipment sent to the Observatorio del Roque de los Muchachos. (Photo: David Calvert, RGO.)

astronomers at British and Netherlands universities and for the United Kingdom Science and Engineering Research Council.

Already on the La Palma observatory site the United Kingdom has built the domes for a 2.5-metre telescope and a 1-metre telescope which, with the 4.2-metre telescope, will provide the full ensemble of telescope sizes necessary at a modern observatory. The 1-metre telescope, of relatively conventional design, will perform survey work for the two large telescopes as well as provide data on the brighter stars in studies which do not need the power of the large telescopes. The 2.5-metre telescope is the Isaac Newton Telescope (INT), formerly at the Royal Greenwich Observatory in Herstmonceux, Sussex. The mirror which arrived in December 1982 is new and replaces the 98-inch pyrex mirror of the old INT with a new glass-ceramic mirror of an almost-metric 100-inch diameter. The glass-ceramic material, Zerodur from West Germany, has a near zero coefficient of expansion, and retains its accurate shape at any temperature and throughout the diurnal temperature cycle of the mountain-top air. The refurbished INT has a modern and accurate control system, a new top-end to support the prime focus camera, re-figured secondary mirrors to match the quality of its new site, and a suite of up-to-date instruments.

Instrumentation for the UK telescopes on La Palma is of prime concern to astronomers at the Royal Greenwich Observatory and at the universities it serves. Broadly speaking, the universe is sending to us now all the information that it ever will about itself. Only with new, efficient and novel instruments will this information ever be grasped and its messages decoded. The INT has a new spectrograph as its principal instrument, in which starlight will be dispersed into a rainbow-like spectrum and recorded. Spectroscopy is the main way that astronomers gain information on the motions and compositions of stars, galaxies, and nebulæ. Two detectors may be used to record the spectra: the Image Photon Counting System (IPCS) and a camera containing a Charge Coupled Device (CCD).

The IPCS is a television camera system equipped with an image intensifier which amplifies light ten million times. Having travelled for 10,000 million years from a distant faint quasar, it is the fate for some individual photons to energize an electron in the sensitive detecting layer (photo-cathode) of this device; the electron is accelerated by a high voltage in progressive amplification. At the final stage a splash of light is produced for each photon detected in the first stage. This splash of light is recorded by the TV camera, recognized as arising from one photon and its position registered in a computer to form part of an accumulating image. In a CCD, by contrast, light falls on to a silicon wafer and frees electrons from it. The electrons are held in place by voltages applied at electrodes on the CCD. After enough light has been received and has freed enough electrons, the voltages on the electrodes are manipulated so as to pass the electric charge off the CCD through amplifiers and to read it into a computer, where it forms the complete image. The CCD is more efficient than the IPCS, and also enables astronomers to inspect the infrared region of the spectrum of quasars and galaxies; but it produces a background of electrical 'noise' which makes it difficult to detect signals from faint stars. The IPCS is virtually free of any background and so is optimized for faint object work. Thus the two detectors are complementary.

Not least of the changes to the INT is its new latitude. At the 51° latitude of Herstmonceux, the polar disk, which gave the telescope its hour–angle motion, lay relatively flat. At the 29° latitude of La Palma, the disk stands more on edge. The edge supports were beefed up and weights were loaded on to the underside of the polar disk to counterbalance the extra leverage of the telescope mounted on the top side. The telescope was erected in the half-finished observatory building on La Palma in the summer of 1981, and the building completed around it by January 1983. It is planned that it should be ready for use by astronomers within a year from then.

The three United Kingdom telescopes on La Palma will be used by astronomers from the British, Dutch, and Spanish communities. The users will not have to be physically present at the telescopes to operate them. Control of telescopes and instruments will be by two Perkin-Elmer minicomputers which will be instructed by a third computer based at the Royal Greenwich Observatory at Herstmonceux. Instructions will be passed by telephone cable or satellite via the commercial international telephone network, for the telescopes to perform an observation, for example, to track a certain quasar and to record its spectrum. These instructions will be issued by an astronomer at the RGO to the Herstmonceux computer. The astronomer will receive by the reverse link information about the position to which the telescope is pointing, the quality of the viewing conditions, and the progress of the observa-

Figure 3. The whole of the scientific installations at the Observatorio del Roque de los Muchachos can be seen in this picture of the rim of the volcanic crater. Right to left the features are: at the highest point the base of the Polaris Trail Telescope used for site testing; the (yet-to-be painted) dome of the 1 m telescope; the white dome of the Isaac Newton Telescope showing over the hill; the Solar Tower belonging to Sweden; huts belonging to the construction company erecting the United Kingdom telescope buildings; triangular derricks stand in the area cleared for the 4.2 m Herschel telescope; and building supplies are stacked next to a cleared area for the Carlsberg Automatic Transit Circle. (Photo: David Calvert, RGO.)

tion, including a representation of the spectrum as it is being recorded. Monitoring these data, the astronomer will choose how to follow up his discoveries and maximize his scientific productivity.

The Herschel Telescope will, like the INT, be a versatile, precision telescope. The 4.2-metre diameter light-collecting mirror is expected to surpass in its optical quality any large-scale terrestrial astronomical mirror, with 90 per cent of the light from a star falling within 0.5 arc sec. This specification of the telescope resolution is designed to match the expected seeing quality of the site. The accuracy of the paraboloidal surface is measured in millionths of an inch, and has to be achieved even though the mirror would deform hundreds of times this amount under its own weight unless supported from below. Support of the mirror is maintained by a flotation system of 64 pneumatic pads. Like the INT, the Herschel Telescope's mirror is made of a glass-ceramic material, this one of CerVit from the USA. The tracking ability of the telescope will match the images produced in the telescope. If, as it follows stars rising and setting in their diurnal motion, the telescope were to waver, the star images which it produces would be noticeably asymmetrical when recorded in the hour-long time-exposures necessary to detect the faintest stars, and this affects measurements of their brightness and position. The amount of tracking error allowed in the Herschel telescope is 0.05 arc sec, equivalent to following the nose on the head of the Queen on a two-centimetre coin at a distance of 4 kilometres, as the coin is carried at 1 km/hr.

Tracking of the telescope will be by an altazimuth mounting, in which the two axes of rotation by which the telescope follows stars

Figure 4. The dome of the Isaac Newton Telescope building on the Roque de los Muchachos in August 1982, nearly completed. (Photo: David Calvert, RGO.)

as they rise and set are horizontal (*alt*itude axis) and vertical (*azi*muth axis). Conventional telescopes (such as the 5-metre Palomar telescope, the 3.9-metre Anglo-Australian telescope and the 2.5-metre Isaac Newton Telescope) are equatorially mounted, or tilted such that the 'vertical' axis is aligned with the polar axis of rotation of the Earth. Such telescopes track stars by rotation about this axis alone. The altazimuth mounting (used for the Soviet 6-metre telescope and the Multi-Mirror Telescope in Arizona) is more complex because motion in two axes has to be synchronized, but because one axis is parallel to the force of gravity, the telescope is symmetrical, more compact, more economical, more rigid, and of higher performance.

The telescope is designed with a *prime focus*, at the focus of the 4.2 m diameter f/2.5 mirror. Although the images of stars on the axis of the parabolic mirror are theoretically perfect, the focal ratio is so fast that aberrations quickly make the off-axis star images exceed the seeing size. A three-element lens has been designed to correct these aberrations and over a ½° diameter (the apparent diameter of the Moon) the telescope will yield images at the prime focus which are smaller than the seeing size.

This means the telescope can photograph a comparatively large area of the sky. This use of the telescope will, however, be comparatively rare, and the camera – holding photographic detectors and triplet corrector lens – will be dismounted and replaced by a second convex mirror which will refocus the starlight to f/11 and direct it back towards the primary mirror to a choice of four foci selected with a switchable flat mirror. One focus, the *Cassegrain focus,* will be through a hole in the primary mirror; most of the time a spectrograph will be mounted here to analyse starlight into its constituent wavelengths. The telescope beam can also be turned by a right angle through the altitude axis of the telescope to one of two *Nasmyth foci* on either side of the telescope. Two platforms fixed here will carry large instruments, such as a high resolution spectrograph to study the light of the brighter stars in greater detail than the lower resolution spectrograph at the Cassegrain focus. Brighter stars can be studied even when the Moon is up, this being a time when fainter stars are lost in the flood of obscuring moonlight over the night sky. When the Moon sets in the course of a night the flat mirror can be used to switch the telescope beam from the high resolution spectrograph at the Nasmyth focus studying bright stars, to the low resolution spectrograph at the

Figure 5. Model of the 4.2 m William Herschel Telescope shows its altazimuth mounting and the Nasmyth platforms on either side of the altitude bearings. (Photo: David Calvert, RGO.)

Cassegrain focus to study faint ones. Thus the use of the telescope can be optimized to the prevailing astronomical conditions, even changes in seeing through the night. British astronomers are presently studying how to implement this so-called *active scheduling* and replace the traditional method by which an astronomer is assigned a particular night and uses it whether or not it results in conditions just right for what he wants to achieve. In this way the La Palma astronomers may be able fully to exploit the 100 or so hours per year when seeing on the Roque de los Muchachos is as good as ½ arc sec.

The universe is a marvellous physics and chemistry laboratory. In studying astronomy the astronomer has no chance of experimenting with the celestial objects; instead he must observe them, looking at celestial experiments which he did not set up under conditions of his choosing. He has to seize the opportunity to study the experiments at the appropriate time and under the most favourable circumstances. Choosing the La Palma site; matching the performance of the telescopes to the site's ideal qualities; optimizing the telescopes' use to the seeing, moonlight, etc; following up discoveries as they are made – these are the design considerations which promise to place the Royal Greenwich Observatory's telescopes at the Observatorio del Roque de los Muchachos amongst the foremost in the world.

Another Ocean in the Solar System?

STEVEN W. SQUYRES and RAY T. REYNOLDS

When the first Voyager images of Europa were received, the initial impression expressed by many was that they looked like the work of an overly imaginative science fiction artist. To those familiar with satellite pictures of the Earth, however, some of the patterns on Europa were startlingly familiar: the highest resolution photos look much like satellite images of Arctic sea ice. This observation by itself is not very significant. Nature has a way of producing superficially similar forms by very dissimilar means. Recent theoretical work suggests, however, that there is in fact a liquid water ocean beneath Europa's icy crust. This view is supported by a more detailed look at the evidence from Voyager.

By now many people are familiar with the intense volcanic activity on Io, the innermost large satellite of Jupiter. The heat source driving the vulcanism is called 'tidal dissipation'. This same heat source operates on Europa as well. Europa's orbit around Jupiter is not perfectly circular. Instead, gravitational tugging by other moons distorts Europa's path slightly, so that its distance from Jupiter varies continually during the course of an orbit. The enormous gravitational attraction of Jupiter distorts Europa's shape, raising a tidal 'bulge'. Because the distance from Jupiter varies, the gravitational attraction and hence the size of the bulge vary as well. With each orbit Europa is stretched back and forth very slightly. This flexing heats Europa, just as a piece of wire is heated by bending it back and forth rapidly. The heating is much less than for Io, but calculations suggest that it is sufficient to maintain most of the H_2O in Europa in a molten state, creating a liquid water ocean several tens of kilometres deep covered by an ice shell that may average as little as a few kilometres in thickness.

This thin ice shell will be very susceptible to fracture, and the evidence in the Voyager images for extensive fracturing is obvious. When a fracture occurs, liquid water may be briefly exposed at Europa's surface. With no atmosphere above it, this water will boil

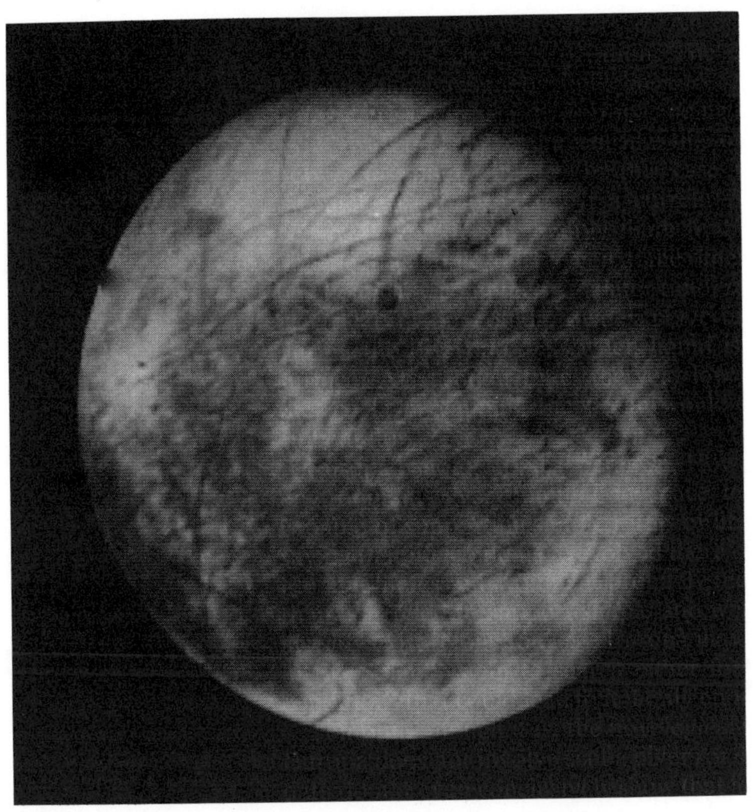

Figure 1. A Voyager 1 image of Europa. (NASA photo).

violently, and at the same time will freeze. There is no paradox here – as boiling occurs and vapour is lost, the vapour carries away with it a great deal of heat. The heat loss is sufficient to freeze a half metre thick layer of ice that shuts off the boiling in only a few minutes. Most of the vapour that boils away is not lost to space, but recondenses on the surface as frost at distances of up to hundreds of kilometres from the original fracture.

There is strong evidence for this type of frost deposit on the surface of Europa. The surface is very bright, and spectra show it to be nearly pure H_2O. It scatters light in the manner expected for a very fluffy frost deposit. The lack of craters on Europa gives evidence for relatively warm sub-surface temperatures that allow glacier-like flow of the crust. Warm temperatures in the crust could be maintained if an insulating blanket of low conductivity frost traps the heat below. Important evidence for frost deposition comes from observations of Europa by the Earth-orbiting International Ultraviolet Explorer spacecraft. These observations reveal that only minute amounts of sulphur are mixed in with Europa's surface frost. Yet electrically charged sulphur originating on Io and trapped in Jupiter's magnetic field is continually bombarding Europa. Unless frost were continually being deposited along with the sulphur, the sulphur would be observed in much higher concentrations. From the observed sulphur concentration we can infer that frost from vapour eruptions is being deposited on the surface of Europa at an average rate of at least ten centimetres per million years.

There is considerable evidence, then, that beneath the ice on Europa lies the Solar System's second ocean. (Second only in order of discovery. It probably existed before the Earth's did.) Its volume, within the limits of our ability to calculate it, is identical to that of the Earth's oceans. Given the known affinity of terrestrial life forms for a watery environment, it is natural to consider the suitability of this proposed ocean as an abode for life.

The temperature and pressure in the ocean should be quite favourable for simple life forms, and it is unlikely that the chemical environment would present any major obstacles. The principle stumbling block is, of course, the lack of a powerful energy source. Only a tiny amount of the sunlight striking the surface of Europa could filter down through cracks in the crust. Yet even this small amount might be sufficient to support very primitive biological activity for limited periods. For example, there are algæ that live at the bottom of permanently ice-covered lakes in Antarctica. During the brief Antarctic summer these algæ receive light that penetrates through the ice, allowing them to photosynthesize so vigorously that they produce oxygen bubbles which actually lift them from the lake bottom. The light levels at which this photosynthesis takes place could exist immediately beneath a fracture on Europa for perhaps five years after fracturing.

Another environment with potential signficance for Europa is found at hydrothermal vents on the Earth's ocean floor. There sulphur compounds produced at very high temperatures are metabolized by bacteria in a process known as chemosynthesis. These bacteria form the base of a very diverse local food chain. If

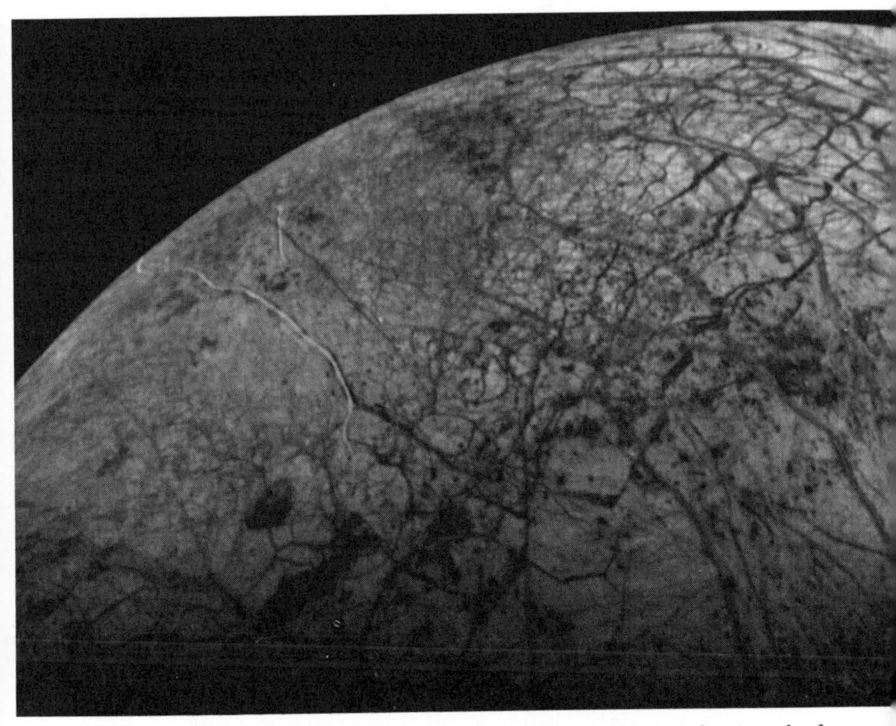

Figure 2. One of the best Voyager 2 images of Europa, showing the network of fractures in the icy crust. (NASA photo).

there is submarine volcanic activity on Europa, then it could be an important energy source there as well. We can calculate the amount of heat that comes from Europa's rocky interior. It is more than that produced by the Earth's Moon, but less than that produced by the Earth. The Earth is, of course, volcanically active, while the Moon presently is not. Europa lies somewhere in between, and we unfortunately cannot accurately predict whether volcanic activity on the ocean floor actually occurs.

Finally, some potentially useful energy could perhaps be stored in Europa's oceans as a result of the moon's motion through Jupiter's intense magnetic field. An ocean on Europa would, like the Earth's oceans, contain many salts, making it a good conductor of electricity. When a conducting object is moved through a magnetic field, one side of the object becomes negatively charged and the other positively charged. An electrical current could be driven by this charge separation. The current would flow from Europa along magnetic field lines to Jupiter, through Jupiter's ionosphere, and back to Europa, being completed in the ocean. Voyager found that a current like this of two million amps flows between Jupiter and Io. In this case the circuit is completed through Io's ionosphere. If a current flowed through Europa's ocean, it would store energy there by bubbling out hydrogen at one electrical pole and oxygen at the other. These gases could then be chemically combined at a later time, releasing their energy. An electrical current could provide a powerful energy source, but only during the brief intervals when liquid is actually exposed at the surface. At all other times the very low electrical conductivity of the ice layer capping the ocean would prevent any appreciable current flow.

Europa is not a tropical paradise. However, it may contain local environments, very limited in space and time, that would be capable of supporting simple forms of life. That is *not* the same as saying that life could have evolved there originally. The creation of life in a lifeless environment probably requires much more energy than is needed to support simple and well-adapted life forms that already exist. None the less, Europa may be one of the best places in the Solar System to search either for life or for the primitive conditions that could lead to it if only enough energy were present.

Tempting though it is, the next logical step in the exploration of Europa does not involve scuba divers or submarines. The Voyager data do not even conclusively show that an ocean exists there; they merely suggest it. The way to prove that an ocean exists would be to actually observe the puff of vapour that would be produced by fracturing of the crust. This might be possible using Earth-based infrared telescopes if the puff were large enough. More likely, though, would be the possibility of actually observing this short-lived cloud of ice crystals and vapour from a spacecraft near Europa at the time of the eruption. This opportunity will be presented when the US space agency's Galileo orbiter arrives at Jupiter in 1988. Galileo will be able to monitor Europa for an extended period of time, and could provide the conclusive evidence that shows whether or not Europa, like the Earth, has an ocean of liquid water.

Author's note. Parts of this article have appeared previously in *The Planetary Report*.

Beyond Pluto

CLYDE W. TOMBAUGH

Within the past two decades, there has been a renewed interest in the origin and evolution of the Solar System. The spectacular success of the deep-space craft in sending detailed pictures of the heavily cratered surfaces of Mercury, Mars, and several of the satellites belonging to the Jupiter and Saturn systems, has contributed immensely to studies of the Solar System. Also, several other items of special equipment aboard the space-craft have made important measurements in the ultraviolet, infrared, radio wavelengths, and polarimetry and spectrography contributing new knowledge regarding the surfaces and atmospheres of the planets and their satellites.

The outer Solar System has come in for its share of attention. My extensive and exhaustive search for other planets after the discovery of Pluto did not reveal any other bodies. If I had searched the Zodiac belt in 1940, instead of 1930, I would have picked up Kowal's mini-planet, Chiron; but, it was far beyond Saturn's orbit in 1930 and about one-half magnitude beyond the limit of my planet-search plates.

For decades after my discovery of Pluto in 1930, there was much controversy whether it was Lowell's predicted Planet X. The mass and size of Pluto had long been uncertain, with various values assigned.

In 1976, Cruikshank, Pilcher, and Morrison of the University of Hawaii, using the 4-metre Mayall telescope on Kitt Peak, found evidence of methane frost on Pluto's surface. This indicated a probable higher albedo and a lesser diameter.

The crowning event occurred in 1977, when Dr James W. Christy of the US Naval Observatory discovered a relatively large satellite to Pluto, very close in. With the measured distance and period of revolution, Kepler's Third Law revealed that the mass of Pluto was only 1/400th that of the Earth. The mean density came out less than that of water ice! What a remarkable sample of the outer Solar System!

In a few more years, the plane of the orbit of Pluto's satellite, Charon, will become edgewise to Earth; and provide transits and eclipses, in which the diameters of both Pluto and Charon can be precisely measured.

In recent years, there have been suggestions regarding the existence of a possible tenth planet, much more massive than Pluto, and further out. But, such a planet did not show up in my 14-year planet search. Presumably, it would be fainter than the 17th magnitude, and the task of finding it would be much more horrendous than what was required to find Pluto.*

* The prospect of the task is described in our book *Out of the Darkness: The Planet Pluto* by Tombaugh and Moore, published in 1980.

The Heart of our Galaxy

DAVID A. ALLEN

Galaxies are conglomerates of stars which, not unnaturally, become more crowded at their centres. If we could journey to the centre of a galaxy, we would see a sky spangled with far more bright stars than is the vault we know. But what else might we find? What lies at the very centre of a galaxy? Do stars crowd so close that they touch? No – for they would rip one another apart, by their mutual gravitational pull. Do they in fact crowd close enough for this to happen? And do other types of object lurk there, in a setting fit for Tolkien tales?

Unfortunately, galaxies are so distant that their central regions become lost in the blurring of our atmosphere. Even telescopes in space will not be able to record in sufficient detail the central cores of nearby galaxies. Our information is deduced by inference. As best we can tell, there are two types of centres to galaxies, rather as chocolates come with soft or hard centres. The norm appears to be an inner region where the concentration of stars is large but not extreme. Here the stars have crowded as close together as they comfortably can. The concentration does not keep on increasing to the very centre, but stays uniform over perhaps as much as several hundred light years, a sizeable fraction of the entire size of the galaxy.

And then there are the abnormal. In these, the concentration keeps on rising inwards as far in as we can tell. The abnormal galaxies, of which very few are known with certainty, have other interesting characteristics. In particular, the best-documented case, Messier 87 *(Figure 1)* has a very intense radio source in its centre, and has ejected a long jet of gas which also emits radio waves.

Now, intense radio sources aren't produced by ordinary stars, and the presence indicates that something much more energetic than starlight exists there. So too does the presence of X-radiation,

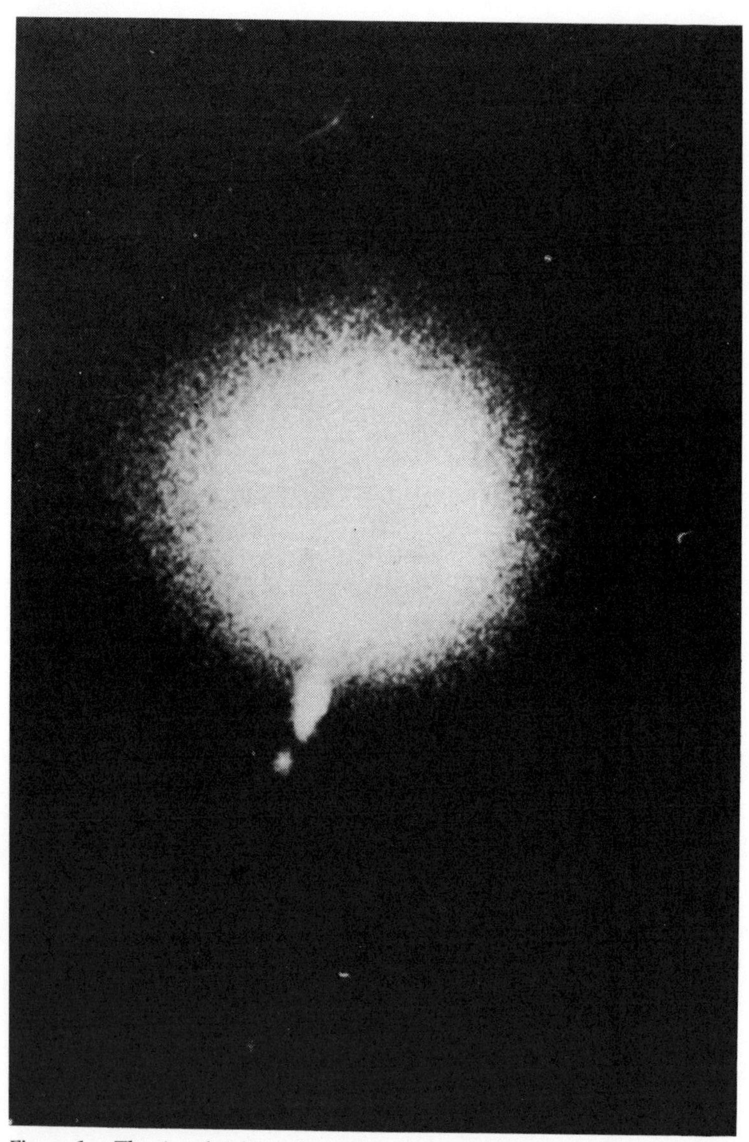

Figure 1. The tiny, bright nucleus of the giant elliptical galaxy Messier 87 emits intense X- and radio-radiation. It is also the origin of the jet of hot gas which stretches far out into the galaxy.

detected by X-ray satellites such as the tremendously successful Einstein satellite. Moreover, when we look at more distant, and thus, more energetic radio-emitting galaxies, we tend to find Seyfert activity in their nuclei.

The trail towards abnormal galaxies has led to the Seyfert phenomenon, and I should digress momentarily to give a definition. I could take an entire page to do so, but in a nutshell Seyfert galaxies have three key features. Something very hot indeed (an X-ray source, in fact) is shining on the gas clouds in the galaxies. Much of that gas is contained in an intensely bright nucleus so small that it looks like a star on photographs. And within that nucleus some of the gas is hurtling around at speeds of thousands of kilometres per second. Think of it: from London to Edinburgh in less than the time it takes a British Rail announcer to say 'XPT'. Quasars, incidentally, are related to Seyfert galaxies: we believe they differ only in being hundreds of times brighter.

Putting all this together, we see that most galaxies have busy, bustling cores, whilst a small percentage feature an extra component, something with a much more bustling core and many other strange properties.

Of course, it makes excellent sense to study the nearest galactic nucleus to us. And that is our own Galaxy, a mere 30,000 light years away. Compare that figure to 2,000,000 light years to the next nearest. Ah – do I hear some of you say? – he's forgotten the Magellanic Clouds! True, they lie only little more distant than our own Galaxy's core, but (sad to say) both are such young, ragged creatures that they have no nuclei.

There is a problem in studying our Galactic centre. It is hidden behind the extensive smog clouds that throng the plane of our Galaxy. So murky are these that between them they absorb rather more of the light than would a well-built brick wall. We must appeal to wavebands other than the visible in our attempt to determine what type of nucleus our Galaxy possesses, and then to learn more about that type of nucleus. Three wavebands are useful: radio, X-ray, and infrared.

The centre of our Galaxy was the first astronomical radio source (apart from the Sun) to be discovered, by Karl Jansky in 1932. Jansky's equipment received radio waves from a large area around the nucleus, and we now know that there are many components to it. Of particular interest is the so-called non-thermal source Sgr A*. The unit of radio intensity has been defined conveniently and named in honour of the radio pioneer; it seems fitting that Sgr A* should have an intensity of one Jansky.

The 'non-thermal' part of Sgr A*'s epithet indicates an abnormal nature, and the source clearly resembles those found in classical radio galaxies and quasars. Radio evidence therefore suggests that we inhabit an 'active' galaxy. The scale of the activity is, however, paltry. We could barely detect the same degree of activity in any other galaxy. The radio source is known to be tiny: little larger than our Solar System which, relative to our Galaxy, is as a pin head to a continent.

Much the same information is gleaned from X-rays. Near the centre of our Galaxy lies a very weak source, probably of the stuff that quasars are made of. The X-radiation is too weak to be mapped in detail: we must assume that the bulk of the radiation emanates from the radio source.

The infrared appearance of our Galaxy's centre is closely akin to the view we would have with our own eyes could the intervening dust clouds be cleared. The reason is that light of long wavelength penetrates dust or fog more easily: the redder the light the better the penetration. At a wavelength of 2.2 micrometres (microns), about four times as long as our eyes see, the dimming of radiation is reduced to a factor of only ten, sufficiently small to allow an almost clear view. Maps of the central regions of our Galaxy have therefore been prominent amongst the work of infrared astronomers. Until recently the best maps were those made by Drs Becklin and Neugebauer (affectionately known in the trade as 'Viking and Fruitcake'), of the California Institute of Technology. It is worth spending a paragraph or so detailing what these maps showed.

There is, as expected, a very large number of stars near the Galactic centre. The brighter ones show individually; fainter stars provide a hazy background whose intensity peaks near the middle, at an elongated feature called IRS 16. The letters stand simply for infrared source; 19 of these were dignified with numbers. Many of the IRS objects are individual stars of very high luminosity – supergiants. IRS 7 is an example. At the shorter infrared wavelengths it is by far the brightest object in the area. Others are more intense at longer infrared wavelengths. IRS 1, 2, and 3 are such. These are clouds of gas and dust heated by local hot stars. The dust glows in the infrared and causes their prominence.

Of the 19 numbered infrared sources, IRS 16 is the one thought most likely to comprise only faint, cool stars. Moreover, it is the closest to that enigmatic non-thermal radio source. Therefore IRS 16 has long been recognized as the counterpart of the centre of our Galaxy.

The Becklin and Neugebauer map, published in 1975, remained the most comprehensive view of our Galaxy's central regions. It had sufficient resolution to distinguish details half of one light year across. By comparison, the smallest detail we could see in any other Galaxy was about 5 light years across.

Technology, of course, advances. When B & N published their infrared map, no X-ray observations had been made, and the radio data were still rather coarse in detail. By the early 1980s it seemed time for a fresh assault on the infrared view of the centre of our Galaxy. That is what this article describes. And whilst the work so far mentioned is purely American, there now enters on the scene the British–Australian axis, *via* the powerful 3.9-metre Anglo-Australian Telescope, the AAT.

In July 1981 I was using the AAT with Harry Hyland and Terry Jones. We were plagued by awful weather in our attempt to improve on the infrared imagery, and as a result we succeeded in making a map at only one wavelength: 2.2 microns, the same as used by B & N. The data were in the form of thousands of numbers written on a magnetic tape by our faithful scribe, the computer. To convert these numbers into the map took many months' work, mostly by Peter Barnes, a Sydney University student who helped us out. Finally we emerged with the view shown in Figure 2.

This image is only 45 seconds of arc across, about the apparent size of the planet Jupiter. Yet within that space we detected hundreds of stars. In Figure 2 the brightest ones have been deliberately over-exposed to show some of the fainter members. The appearance is very similar to a globular cluster, with a vast concentration of stars in the very central part. It is worth stressing that only two of the stars in Figure 2 can be seen on deep optical photographs – and these two are among the very faintest on the infrared image.

It is possible to learn a few interesting facts from this image. One is, that compared to other parts of the Galaxy, there is a paucity of the very brightest sources, which probably means that the stars are extremely old so that all the biggest (and hence brightest) have already departed this life as supernovæ. Also, the

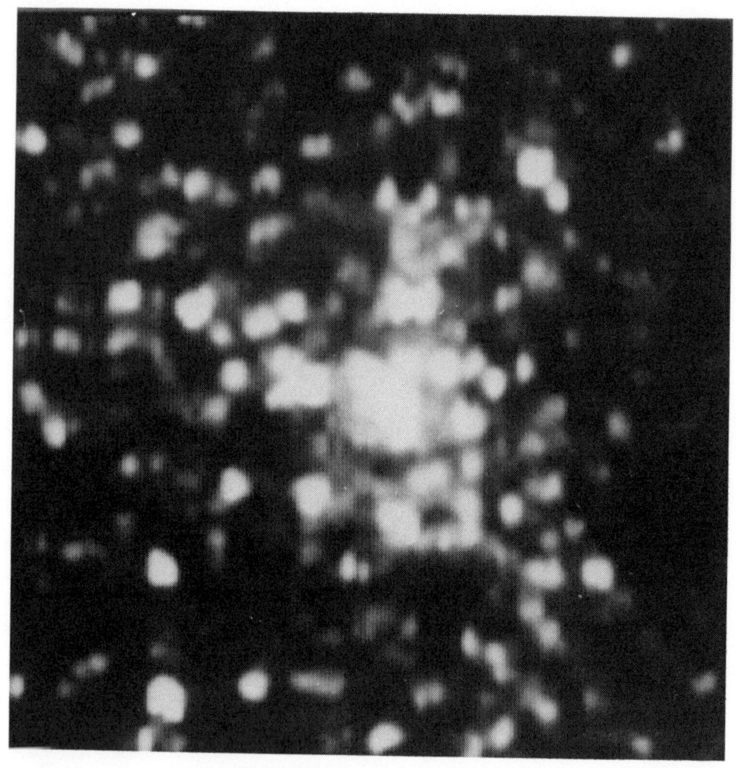

Figure 2. If we could see through the dust of our Galaxy, this spectacularly rich concentration of stars at its centre would greet our eyes. This picture was made in infrared light at a wavelength of 2.2 microns.

image shows barely any hint of a limit to the star density; right in to IRS 16 the concentration of stars is still increasing. Finally, IRS 16 is not a single object at all, but three sources.

This piece of work focuses our attention more on IRS 16. What are the three components? Which (if any) is the true nucleus of our Galaxy? And is it made only of stars, or of something more esoteric?

At this juncture, a red herring turned up to set us back a little.

We acquired a new instrument.

Jasper Wall built it, at the Royal Greenwich Observatory, with the able assistance of Paul Jorden and David Thorne. He brought it to the AAT in 1981. On its first telescope run, this CCD camera was turned towards the centre of our Galaxy. And, like all electronic cameras, this one probed fainter, and more into the red, than any normal photograph could hope to do. Electronics are, indeed, gradually displacing the photographic plate in astronomy. But that's another tale.

It was John Storey who suggested making an exposure of the Galactic centre with the CCD camera. John is a lovable character, tiny of stature but big of heart, and sporting a wry wit. When the exposure was made and appeared on a television monitor, John was delighted to find two stars right at the Galactic centre, stars that had never before been seen. These two appeared to lie almost exactly at the ends of IRS 16, a most interesting location indeed.

The discovery first appeared in print on the front page of the *Coonabarabran Times*, a local newsrag with a circulation of a scant few thousand. The photograph was actually used to advertise the forthcoming Observatory open day. However, from western New South Wales, the discovery spread to the pages of *New Scientist*, where it was interpreted as a 'double mini quasar' at the centre of our Galaxy. A wild conjecture, as time was to prove.

Some months later, John attended a conference in California and learnt that two American groups had also discovered the two stars, using their own CCD detectors. It was time to make another move towards unravelling the Galactic centre.

John and I gave some thought to these new objects, and decided on an approach during the next observing season that would disentangle more fully the activity at the Galaxy's heart. Our proposal was to use three different instruments, in five different ways. We made our case for telescope time, and were granted it. Miraculously the weather gods smiled on our attempt. Considering their scientific mien, astronomers are strangely superstitious about weather. Both John and I are credited with good luck, so perhaps it was no surprise that we had not only clear skies, but crisp and steady ones too. And from Australia we also had good access to that part of the sky. The Galactic centre lies in Sagittarius, and passes almost overhead during the long winter nights.

Our first task was to improve on the CCD images, by observing at several wavelengths and for longer. The new images we acquired amply confirmed the two stars, and gave a suspicion of two others, very close by. But disappointingly, the various wavelengths we used demonstrated that the stars were not nearly so red as they would have to be if seen through the amount of dust that overlies the Galactic centre. We had to conclude that these stars lay only part way to the Galaxy's heart, and therefore were of no relevance.

One chance remained. If really quasar-like, the heated gas in these objects would cause them to have peculiar colours, and this might explain their embarrassing blueness. So we put the CCD camera on to a spectrograph and found out what their light was made up of. No hint of hot gas showed up – the two objects were, as best we could tell, mundane stars.

And so we reverted to the infrared again. Jeremy Bailey had recently put together a lovely bit of computer software which enabled one to scan the telescope around in synchronism with the infrared detector, and so build up a picture of the patch of sky as the data came in. No more laborious disentangling of the numbers on a magnetic tape; no more waiting for months to see the results; and higher accuracy too. We used this.

How dramatically different from the CCD view at a wavelength of 10,000 Å is the infrared images at 12,000 Å! On the latter the CCD stars are barely seen, but four new objects are bright and prominent. Three of these, clearly distinct objects, are the components of IRS 16.

The two brightest infrared sources correspond to the two faint objects on the CCD image, and so at last we know where the optical and infrared images mesh. And how cruel is chance: the stars seen on the CCD exposure, those two red herrings that were not red enough, parallel the ends of IRS 16 but lie less than two seconds of arc to the north. Two seconds of arc projects to 50 metres on the Earth's surface, a minute detail if you are looking at a map of London and its suburbs. No wonder the *New Scientist* reporter (and others) jumped to the conclusion that those CCD stars were intimately related to the Galaxy's heart.

We made infrared images in great detail of the central portion of the area covered in Figure 2. Two of these are shown in Figure 3, and illustrate how dramatically different the appearance can be at different wavelengths. These images also demonstrated that all other sources in the vicinity are redder than IRS 16's components. We take this to mean that IRS 16 contains the hottest stars.

By very careful measurement of positions, John and I were able to show that the non-thermal radio source lies close to the central component of IRS 16. We believe that the two are coincident, and

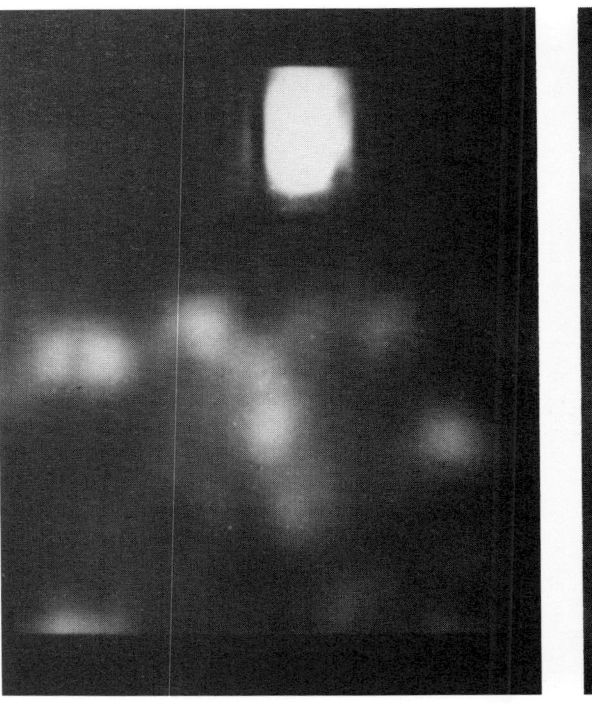

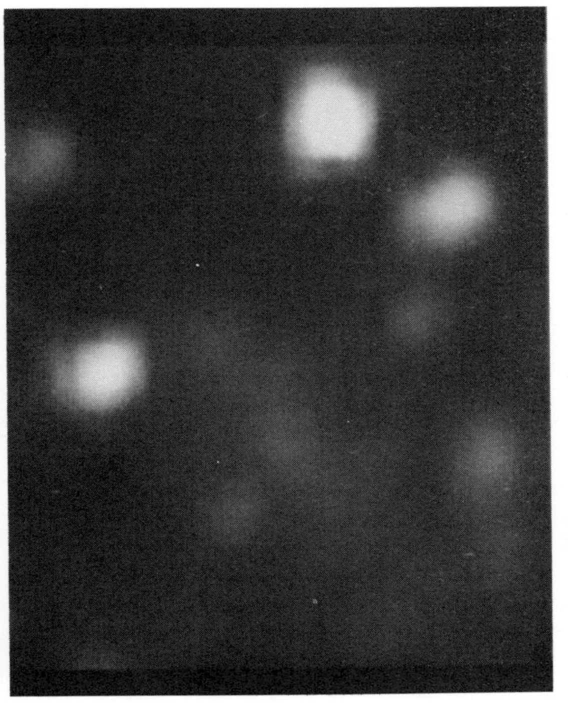

Figure 3a. *Figure 3b.*

Detail of the innermost regions of the Galaxy at 2.2 microns (left) and 3.8 microns (right). The bright source at the top is the very luminous single star which is distorted in Figure 2 to appear double.

that this tiny concentration of giant stars is the Galaxy's very centre. But if so, then what are its immediate neighbours, the sources to upper left and lower right of the triple structure in the middle of Figure 3a?

To answer this question, we finally resorted to spectroscopy in the infrared. We examined two chemicals available for study at these wavelengths, carbon monoxide and heated hydrogen gas. The CO arises in the atmosphere of cool stars, and our observations confirmed that the right sort of stars are present in the middle source of IRS 16 but not in the other two components, where the stars are hotter. The hydrogen observations showed that there is little or none of this gas in the very central region, but that hydrogen lies at the extremities of the IRS 16 complex. Very likely it is here heated by the embedded hot stars.

The picture is yet incomplete, though we do now have a much clearer view. It looks as though IRS 16's central source is the true core of our Galaxy, a region dominated by a host of cool stars and certainly not Seyfert-like. If so violent a creature as a black hole lurks there, it is currently dormant. This is quite in order, for some people argue that Seyfert activity is an occasional and recurrent feature of many galaxies. Possibly black holes gather up gas lost by stars. In so doing they will radiate strongly. This luminosity blasts away the remaining gas and leads to a quiet spell. We may see our Galaxy now in such a dormant state. This is, of course, only speculation. If correct, then Fred Hoyle, in his science fiction book *The Inferno*, was not so far off the mark after all.

Diary of an Observational Astronomer

MARTIN COHEN

In front of me, a wide-bodied jet is pushed away from the air terminal, its streaked fuselage the only colour in the otherwise grey morning landscape. I am heading for Mauna Kea, the volcanic peak of Hawaii's Big Island, and at the end of my journey is a 150-inch telescope. Modern astronomers are a highly mobile population, often flying thousands of miles between their home institutions and the observatories at which they work. Come with me on this present trip and see why I am visiting Hawaii, the equipment I shall be using, and the science that I hope will be furthered by these observations.

My first professional work in astronomy was in the infrared region – a vast chunk of the spectrum extending roughly from the limiting red response of photographic plates (about 1 micron, or ten thousand Ångströms, in wavelength) to the microwave radio realm (around one millimetre). Infrared astronomy is still a relatively young science, not yet out of its second decade. Its practice makes special demands upon a site and the design of a telescope. First, as in all ground-based astronomy, the Earth's atmosphere intervenes. There is a host of gases that absorb selectively and strongly throughout the infrared. Consequently we operate in the narrow windows between these often totally opaque absorption features. Most greedy of these gases is water vapour. By careful selection of a high, dry, mountain site the water absorptions can be minimized and their time variations reduced.

Secondly, infrared can loosely be regarded as heat; that is, any warm body emits thermal radiation in accordance with its temperature and physical properties. The hotter the object, the more intense is this emission, and the shorter the wavelength at which the spectrum of heat radiated will achieve its peak. Since our atmosphere is not totally transparent, some radiation is absorbed, and this serves to warm the gases. Terrestrial atmospheric emission unfortunately peaks in a rather broad and astrophysically useful window, creating a bright background near ten microns in wavelength. These considerations explain why in the infrared there are no 'night' and 'day', and why infrared astronomy can be likened to looking for warm needles in a hot haystack.

The technique of sky chopping was evolved to cope with this powerful background. Two small beams are defined on the sky, and the infrared sensor is rapidly switched from one to the other, comparing the two signals. One beam includes sky and an object of interest; the second only bright, nearby sky emission. By subtracting one signal from the other, the sky signals are largely eliminated. Of course, the telescope and its optics are also warm and produce their own background. However, careful optical design together with chopping can minimize this internal background.

In short, we need a custom-designed telescope, located in a high, dry site which must also satisfy the usual astronomical constraints of cloudless and visually dark skies, and good atmospheric 'seeing' (that is, minimal turbulence that distorts and moves the final images). One of the world's best infrared sites is atop volcanic Mauna Kea, almost 14,000 feet above the Pacific Ocean. Indeed, in the past few years, international enthusiasm has settled no less than three major new telescopes on the peak, none less than 120 inches in aperture. Canada and France collaborated on a multipurpose instrument; the United Kingdom erected a giant 150-inch infrared telescope; and NASA constructed a 120-inch infrared facility. Mauna Kea Observatory is a very special place!

The Pacific is covered by low clouds as we rise through hazy conditions into a clear blue world and I begin to check my sheaf of observing notes and charts. Some years ago, while at Berkeley, I and a friend became interested in the infrared properties of hot stars; objects much more massive than our Sun, with surface temperatures in the tens of thousands. We found evidence in some stars for unexpectedly large quantities of infrared radiation, far in excess of the simplest predictions based on stellar models. The signature of these excesses is low temperature emission often in the range 100–300 degrees absolute. As the clues mounted from studying different categories of hot star we realized that we were witnessing heat emitted by cool circumstellar dust grains. The dust absorbs a little of the ultraviolet and optical starlight, becomes heated, and reradiates this energy in the infrared. One mystery was the origin of the dust. So powerful are the winds blowing out of the particular stars that we had observed, that theoretical considerations indicated hypothetical grains would be essentially pounded into oblivion within only hundreds of years by the rapid flows of gas. Consequently, we were led to the conclusion that the grains were actually forming within the speedy outflows, moving with them, thereby greatly reducing their scouring action.

Exciting though this was, not all of our hot stars showed this type of cool dust radiation. Other stars displayed much smaller excesses at the longest wavelengths. We re-observed several of these and decided that the phenomenon responsible for their anomalous emission was a purely gaseous one. At these high stellar temperatures atoms of many species are ionized, that is, deprived of some electrons. When this accumulated sea of electrons washes past heavy ions, radiation is produced. This we term 'free-free' emission, since electrons are not tied to specific heavy ions either before or after emitting their characteristic radiation.

Eventually outflowing gas forms essentially an extended atmosphere around the star. My colleague, an adept theoretician, realized that, given an adequate model for the distribution and velocity of the material, a measurement of the infrared excess radiation would yield an estimate of the rate of loss of matter from the star. There is great interest currently in the nature of hot stellar winds. By comparison with our Sun, massive stars are returning gas to the interstellar medium at prodigious rates that may attain to a billion times that of the solar wind for an individual star. Another reason for the growth of the topic is that ultraviolet astronomy also has recently been born. Bright glowing emission lines characterize the ultraviolet spectra of these hot stars. Spectroscopic analysis of these lines offers vital clues to the acceleration process of the winds and provides quantitative estimates of the velocities to which gas is finally driven.

Our study is one for which we feel great enthusiasm, partly because it demonstrates that in modern astronomy one must bring to bear on specific problems every kind of technique, both observational and theoretical. We use optical data on our stars to predict the normal amount of infrared energy, in the absence of free-free emission. Infrared measurements then yield the excess radiation, and the crucial issue here is to acquire data of high enough quality to determine the often small excesses with precision, above statistical uncertainties in the observations. To translate from infrared data to mass loss rates further requires theoretical modelling, based upon velocities derived from the ultraviolet spectra. Radio-astronomical instrumentation is now achieving a performance whereby the weak level of free-free radiation emitted at centimetre wavelengths by the most luminous stars also is detectable. Radio observations provide further checks on the theory, and so far there is good agreement between mass loss rates determined independently from infrared and from radio data.

Hilo is hot and humid as our plane lands. We are met by friends whom we have not seen for several years, and are shuttled to the headquarters of the UK Infrared Telescope Unit. Here we meet the staff of mechanical and electronic engineers, secretaries, receptionists, astronomers who provide support for visitors to the telescope as well as pursue their own research programmes. Quickly we are brought up to date on the performance of the available infrared sensors. We discuss the infrared filters we require, and which sensor to use for each of the several nights on the telescope. Soon it is time to try to overcome our jet-lag and the long first day of the trip ends.

On the next day we trade ideas and information on our own research with the staff scientists, have a brief sea-level medical for subsequent comparison with our performance at high altitude (more than one type of scientific research can be carried out at an observatory), and use the computer to update coordinates of our programme stars to allow for precession. Later, several of us pile into a truck with four-wheel-drive and ascend to the mid-level facility at 9000 feet on Mauna Kea. This will be our home for the next few days. Astronomers from the several telescopes and institutions represented on the peak live, eat, and relax here and drive to their respective domes each evening. On the drive we have noted an abrupt transition from dense, wet cloud to clarity a little below the altitude of the dormitory site. Above this temperature inversion the atmosphere should be clear and dry. There is a lot of moonlight as we stare up at the sky, trying to re-orient ourselves to the low latitude of twenty degrees north. We can see a few shreds of wispy cirrus, white against the darkish sky, and we inwardly hope that our own nights will not be fraught by these often slow-moving, high-altitude clouds. Clouds not only attenuate celestial infrared signals, they also emit by virtue of their own heat content. In short they are an annoying source of extra noise and their presence severely limits the type of infrared work that can be pursued on a given night. We locate our dormitory rooms and gratefully get into bed, beginning the process of acclimatization to altitude.

During the succeeding days we fall into a life-style in which day of the week and date lose all meaning. Life is circumscribed by a routine. We awaken between two and four p.m., eat dinner at five, and then body-shakingly ascend the more or less graded rocky dirt road that snakes its way to the summit of Mauna Kea past cinder cones and jagged lava. The sun has just set as we open our dome to release trapped warm air that would jeopardize the internal 'seeing' or low-level atmospheric steadiness. We first acquire a bright star to use for infrared calibration and to establish the pointing of our detector relative to the television camera in the telescope which takes the place of an observer's eye. There is no need to sit in the cold, darkened dome for this project. Almost all the necessary functions can be triggered remotely from the warm, lighted control room where computers hum and look after telescope slewing and tracking, and display and digest our infrared signals. After twenty minutes we are content with the focus and alignment of the system, and with the infrared signals received from our standard star. We type in the coordinates of our programme stars, acquire them on the television screen, and start to collect infrared data. A chart recorder enables us to maintain a continuous watch on sky conditions and on our signals without leaving the control room. Should the atmosphere become moist, and clouds begin to form over us or drift in at high altitude, the noise on the chart recorder at certain wavelengths will become gruesomely large, alerting us to the danger. Or, we may be lucky, and have to contend with only modest 'sky noise' throughout the night.

Once the instrumentation is set up for the night, only the weather can halt the proceedings. Our *modus operandi* is first to measure objects at short infrared wavelengths; then to examine the signal at the longest wavelengths. It is always exciting to watch the first few numbers come in at the computer, especially if the signal is unexpectedly large. What physical reason could possibly cause a star to be losing mass at such a high rate? How do our infrared data compare with the radio signals? Could one have guessed that this star would be so interesting from its optical characteristics alone? We proceed slowly down our list of priority objects. All too soon it is dawn, and the television begins to object to the bright skies. We put the telescope to bed, and bounce dustily down the mountain to our own beds, watching the sky anxiously for clues to the weather for the next night.

After ten days it is time to return home, bearing whatever data we have been lucky enough to glean from the sky, and to begin the work of interpretation and modelling. But for the observer, there is nothing to compare with the thrills of using a large telescope efficiently, of seeing infrared signals for the first time from a previously unobserved object. For me the actual process of observation is an important part of my own enjoyment of astronomy, and to be able to use the world's largest dedicated infrared telescope is a great privilege.

Probing the Galaxy with RR Lyræ Stars

F. G. WATSON

Many classes of astronomical object are as interesting for what they tell us about their environment as they are in themselves. History is full of examples of astronomers capitalizing on the occurrence of such useful objects and, indeed, some of the most fundamental advances in our understanding of the Universe have taken place as a result. Kepler 'used' the planets in formulating his laws of planetary motion, thus setting the stage for Newton; Shapley 'used' the globular clusters to discover the Sun's non-central position in our Galaxy; Hubble 'used' the galaxies themselves to deduce that the Universe is expanding, and so on. And the process continues today, with astronomers employing a whole variety of objects as test-particles in a variety of situations.

Some of the most important things we have learned about the structure of our own Galaxy have come from observations of stars. In this article I should like to describe how we have been aided by one very useful group of stars – the RR Lyræ variables. In particular, I want to highlight some research using RR Lyræs, currently being carried out at the Royal Observatory, Edinburgh, which could significantly affect our understanding of the internal behaviour of the Galaxy.

But, first of all, what *are* RR Lyræ stars? The short answer is that they are pulsating giants with periods of less than one day, although that bald statement gives little hint of the enormous amount of theoretical and observational study that has been devoted to them over the years. RR Lyræs were first observed in the globular cluster Omega Centauri by S. I. Bailey of Harvard Observatory in 1895, and it was not until four years later that the eighth-magnitude star which eventually gave the class its name was discovered. Bailey divided his 'cluster variables' into three subclasses – conveniently denoted a, b, and c – with differing light-curves, periods, and amplitudes. In fact we now know that c-type RR Lyræs differ from the rather similar a and b types in

Figure 1. Typical of its class, the globular cluster NGC 362 is a spherical aggregation of more than a hundred thousand stars. It was in systems like this that RR Lyræ variables were first observed. (Photograph courtesy UK Schmidt Telescope Unit, © Royal Observatory, Edinburgh)

that they are oscillating in an overtone mode rather than the fundamental – acoustical terms which readily lend themselves to the science of stellar pulsation. Because of this, the properties of c-types are somewhat poorly understood and they are, therefore, much less useful for the kind of galactic probing that we wish to do.

All the observable properties of an RR Lyræ star – not merely its brightness – vary as it goes through its pulsation cycle, and sometimes the variations are extremely complex. A few of the more well-behaved characteristics of a typical a-type RR Lyræ are shown schematically in Figure 2. The star will have a period of about 0.45 days (it would be rather longer for a b-type) and its visible light amplitude will be some 1.5 magnitudes (less for a 'b'). The most striking property of the light-curve is its saw-tooth appearance, very similar to that of a classical Cepheid variable, to which the RR Lyræs are distantly related. It is closely mirrored by the radial-velocity curve beneath, which shows velocity changes in the star's reversing layer, or outer atmosphere, as it pulsates. Note that the star is at maximum brightness when its surface is approaching the observer at maximum velocity, and that the velocity amplitude can easily reach 100 km/s.

A relatively straightforward calculation enables us to use the velocity curve to plot the actual change in the star's radius throughout its pulsation cycle (Figure 2(c)). The variation may amount to 10 or 20 per cent of the star's mean radius, and is a major contributor to the change in its luminosity. At visible wavelengths, this combines with variations in the star's temperature to produce the characteristic saw-tooth light-curve we have already seen. But at infrared wavelengths, we should expect the temperature variation to play a less important role (because we are observing a flatter portion of the black-body spectrum) and should therefore see a radius-dominated light-curve. Indeed, that is exactly what we do see; some very recent observations, made with the 3.8-metre UK Infrared Telescope in Hawaii by a group of astronomers (including this one) led by Andy Longmore, produced light-curves of the kind shown in Figure 2(d).

RR Lyræ stars exhibit some of their most unruly behaviour in their spectra. For a start, most RR Lyræs defy normal classification in the spectral sequence O, B, A, F, G, K, M. The problem is that the intensities of their hydrogen absorption lines and their metal, or calcium, absorption lines do not correspond to a unique spectral type. Thus, while both sets of lines will yield classifications typically in the range A0 to F6, the two may differ by as much as a spectral class – for example, a calcium spectrum of A5 appearing with a hydrogen spectrum of F5. To make matters worse, both these spectral types will vary quite dramatically throughout the pulsation cycle of the star. It took a major study by Lick Observatory astronomer George Preston in the late 1950s to make some sense of the situation. He defined a spectral index, $\triangle S$, to be the difference at minimum light between the hydrogen spectrum and the calcium spectrum in units of one-tenth of a spectral class. Thus, the example mentioned above would correspond to a $\triangle S$ of

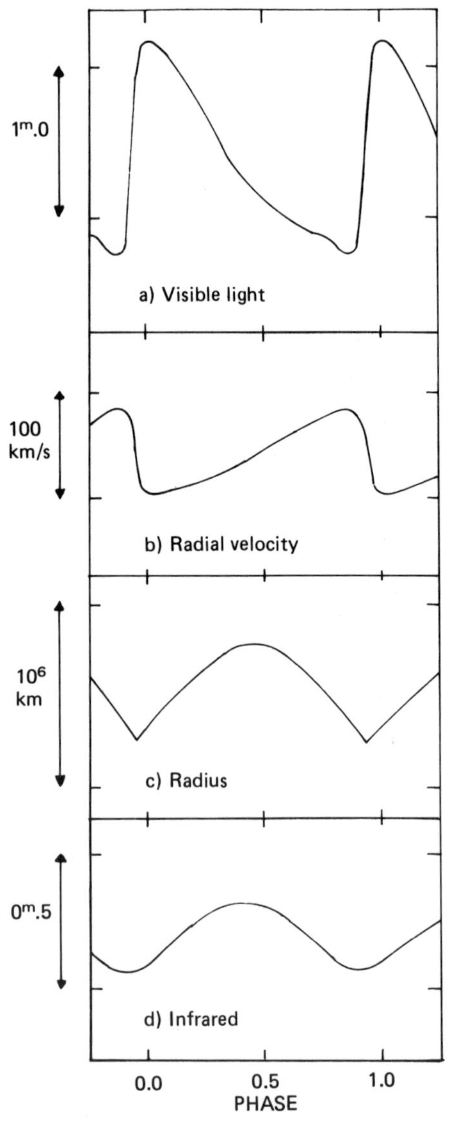

Figure 2. Changes in some of the properties of an a-type RR Lyræ star throughout its pulsation cycle (see text). Phase is measured from visible-light maximum, in fractions of a period.

10. Preston's index is normally taken as an indicator of the 'metallicity' or, more correctly, the 'metal deficiency' of an RR Lyræ in the sense that a star with $\Delta S = 0$ would be considered to have a roughly normal mixture of chemical elements (the so-called 'solar abundance') whereas one with $\Delta S = 10$ would be very metal-weak. This immediately suggests one possible use for the RR Lyræ stars – as tracers of galactic chemistry – and, as we shall see, they have indeed been employed as such.

What of the occurrence of RR Lyræ stars? With the exception of the long-period variables, they are the most numerous of all intrinsic variable stars and, in our Galaxy, may outnumber their relatives, the Cepheids, by more than six to one. It is important to realize that in spite of the similarity of their light curves, and a common pulsation mechanism (instability due to the second ionization of helium at a particular depth in the star's atmosphere), the Cepheids and the RR Lyræs form two quite distinct classes of object; the RR Lyræs are not merely faint, short-period Cepheids. This is underlined by the fact that we find them in different places in the Galaxy, the classical Cepheids being strictly confined to the disk whilst RR Lyræs turn up in the halo, the nucleus and, of course, the globular clusters – as well as the disk. These distinctions, and others, lead us to regard the two as Population I and Population II objects respectively.

The attribute of RR Lyræ stars that makes them most useful to us is, paradoxically, their most hotly debated parameter. Theory predicts that RR Lyræs should all have rather similar absolute magnitudes (i.e. magnitudes as seen from a standard distance) with, perhaps, variations correlated with period, colour, metallicity, and so on. By and large, this is borne out by observations, in that RR Lyræs at the same distance (or nearly so), as in a globular cluster, all have similar *apparent* magnitudes. But if we are to use the RR Lyræs as distance probes in the Galaxy, then we wish to know their absolute magnitudes as precisely as possible. An uncertainty in absolute magnitude of only $0^m.5$ would translate itself to an uncertainty in distance of 2 kiloparsecs, or 6,500 light-years, at the galactic centre! So, a good deal of very careful work has gone into the determination of RR Lyræ absolute magnitudes – and it has met with problems.

For example, the predicted variations of luminosity with period, etc., that I mentioned, do not always seem to behave as expected. Neither do the brightnesses of RR Lyræs from different parts of the Galaxy when intercompared, and it now seems that some of the RR Lyræs in the immediate neighbourhood of the Sun are systematically brighter and less metal-weak than their cousins in the halo. But the biggest problem undoubtedly lies in calibrating the zero-point of the absolute magnitude scale, simply because there are no RR Lyræs whose distances are small enough to be measured directly. Consequently, investigators have had to resort to a variety of statistical methods, from which have emerged, naturally, a variety of solutions of differing significance and interpretation. Fashions change, too, and whereas ten years ago a value for $<M_V>_{RR}$ (the absolute visual magnitude of an a- or b-type RR Lyræ averaged over the light-cycle) of $+0^m.5$ would have seemed reasonable, the trend is now towards somewhat fainter values. A recent and very extensive study by Victor Clube and John Dawe of the Royal Observatory, Edinburgh, has resulted in the unusually low value of $<M_V>_{RR} = +1^m.0$ (for $\triangle S=7$) with some evidence for variations with period and metallicity. Working not a thousand miles from these astronomers (in fact, in close collaboration with them), I have adopted this value in the studies that I have carried out. Of course, any future readjustment of the value of $<M_V>_{RR}$ will not necessarily be fatal to existing research, because many results can simply be scaled accordingly. For this reason, distance scales (e.g., of globular clusters) are often quoted in terms of the adopted absolute magnitude of RR Lyræ stars.

It is unlikely that the adopted value of $<M_V>_{RR}$ will be pushed any fainter than $+1^m.0$ because of the consequences in a rather larger context. RR Lyræ stars play an important role in determining the distance scale of the Universe at large, since they can be observed in a few nearby galaxies. Making them as faint as $+1^m.0$ has the effect of reducing our estimate of the distances to these galaxies. This, in turn, renders the Universe smaller than hitherto supposed, and hence younger, because of the shorter time since the big bang – in fact, only a little over 10^{10} years. Some of the oldest objects in our Galaxy are globular clusters (about the same age – 10^{10} years), and many contain RR Lyræs which seem to be intrinsically brighter than the $+1^m.0$ of those in the general field, perhaps because they are less metal-weak. This suggests (see later) that the faint RR Lyræs in the field could actually be older than the globular clusters. To make them fainter still would further increase their ages and simultaneously reduce the age of the Universe, so that a situation would soon be reached where the RR

Lyræs were older than the Universe itself – clearly an unlikely state of affairs! Actually, it is arguments rather like these that lead some astronomers to adopt a steady-state model of the Universe in preference to the big-bang theory widely accepted today.

There is one rather interesting development in this area of study now on the horizon. 1986 will see the launch of ESA's high-precision astrometry satellite, HIPPARCOS, which, with its milli-arc-second accuracy, might just be capable of directly measuring the distances to the nearest RR Lyræs with marginally better precision than we think we know them already. If nothing else, it will provide a few additional data points for the RR Lyræ distance-scale pundits to argue over!

The structure of the galactic halo

The halo, or corona, of our Galaxy, is the spheroidal aggregation of globular clusters and stars that surrounds the disk. It stretches outwards from the bulge around the galactic nucleus, gradually falling off in density until the emptiness of intergalactic space is reached. As far as galactic explorers like ourselves are concerned, it offers the enormous advantage over the disk of containing little, if any, gas and dust, and is consequently almost transparent. Lines of sight in the halo tend, therefore, to be long, unlike the very restricted views we have in the murky spiral arms of the disk. The stellar component of the halo is important; it is suspected that there might be very large numbers of low-luminosity stars which make up a so-called 'missing mass' in the Galaxy. However, the RR Lyræ stars, which also populate the halo, are *not* low-luminosity objects and, indeed, shine like beacons, delineating the galactic halo out to very large distances.

By making photometric observations of RR Lyræ stars we can learn a very great deal about the large-scale geography of the galactic halo. The basic method is to probe the density in different parts of the halo by counting the numbers of stars we find there. This necessitates knowing how far away the stars are, and it is because the distances of RR Lyræs can be calculated reasonably satisfactorily that they have proved the best tools for the job.

The apparent magnitude, m, absolute magnitude, M, and distance, d (in parsecs, units of 3.26 light-years), of any celestial object are related by the formula

$$m - M = A - 5 + 5 \log d$$

where $m - M$ is called the distance modulus. The quantity A is the number of magnitudes of dimming, or absorption, caused by material between the object and the Earth; it can be estimated by indirect means. It would, of course, be zero if the intervening space were perfectly transparent, but this is seldom the case, even in the halo. Now, in spite of all the problems I have just described, RR Lyræ absolute magnitudes can be estimated with more certainty than for almost any other type of star. So, by measuring their apparent magnitudes very precisely, we can get distances accurate to ten or twenty per cent, with possible allowance for a scaling factor of the kind already mentioned.

Having decided that we will be able to measure the distances of our RR Lyræs, we next have to find them. It is for this purpose that a number of surveys for RR Lyræ stars have been carried out, in various strategic directions where we are most interested in sampling the density of the galactic halo. The variables are found by taking a large number of photographic plates of the selected star-field and intercomparing them by some means, usually a blink microscope. The RR Lyræs reveal themselves by their rapid variation, and then periods and magnitudes can be determined. Unfortunately, the directions in which we would most like to observe are often rather close to the disk of our Galaxy, and this means that our view is obscured, not by material in the halo, but by nearby gas and dust belonging to the spiral arms in which the Sun is embedded. This obscuration is not uniform, though, and here and there we find holes, or windows, which allow us to penetrate into the halo. It is the existence of these that has determined the particular lines of sight along which RR Lyræ surveys have been undertaken, and the best-known are shown on the schematic cross-section of the Galaxy in Figure 3.

The directions away from the galactic nucleus (anticentre direction) and perpendicular to the disk (north galactic pole direction) were searched for RR Lyræs by Hoffmeister and Kinman respectively, during the 1960s, in order to place limits on the overall size of the halo. Both surveys revealed RR Lyræs at distances in the region of 50 kiloparsecs (160,000 light-years), indicating that the halo is very large indeed, and much larger than the disk. In addition, the galactic pole survey showed the fall-off in density away from the plane of the disk, and thus gave information on the magnitude of the gravitational force towards the disk. Of course, the south galactic pole (i.e., ‘downward’) direction also offers the possibility of an RR Lyræ survey and, in fact, large

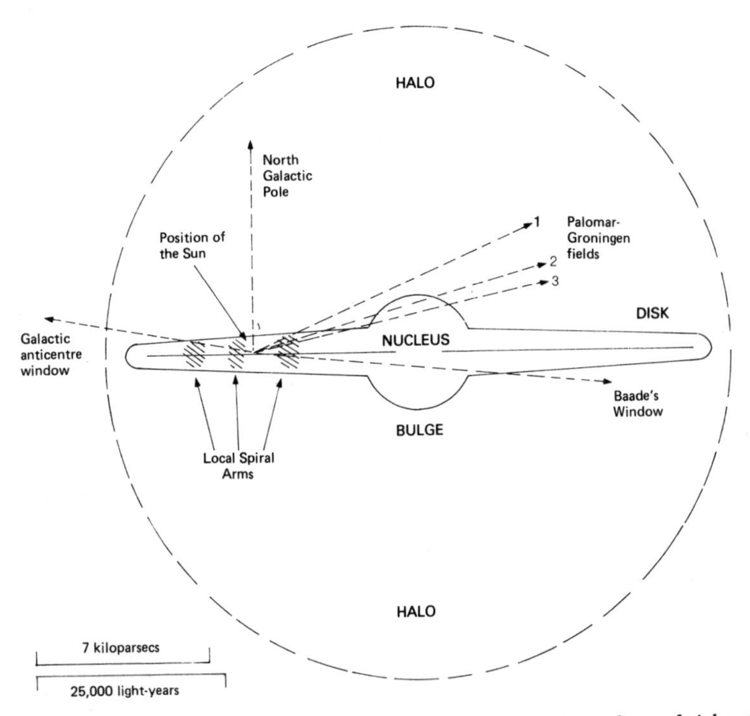

Figure 3. Cross-section of our Galaxy showing the approximate lines of sight of important RR Lyræ fields (dashed lines). The three Palomar-Gröningen fields are all shown north of the galactic centre for clarity; in fact field 3 is to the south. The positions of the three spiral arms near the Sun are inferred from young stars and gas clouds in the disk. They are usually referred to (going inwards) as the Perseus arm, the Orion arm, and the Sagittarius arm.

areas of the sky around both galactic poles are sufficiently free of obscuration for searches to be carried out there too. Photographs of one such region, taken with the 1.2-metre UK Schmidt Telescope in Australia, are currently being analyzed for variable objects by Edinburgh astronomer Mike Hawkins with the Royal Observatory's COSMOS high-speed automatic plate measuring machine.

For most astronomers, the really interesting lines of sight are those towards the galactic nucleus. Figure 3 shows the principal ones, although new surveys are currently being carried out by groups at Cambridge (again using the UK Schmidt Telescope), at

Leiden (using European Southern Observatory Schmidt plates), at the Cerro Tololo Interamerican Observatory in Chile, and at Palomar with the 48-inch Schmidt. The first major galactic centre survey was carried out in a classic piece of work by Walter Baade with the Mount Wilson 100-inch telescope during the 1940s. He found an area of sky, now known as Baade's Window, whose line of sight passes only 4° below the galactic centre, and which contains more than a hundred RR Lyræ stars. When Baade plotted the number distribution of these stars with apparent magnitude

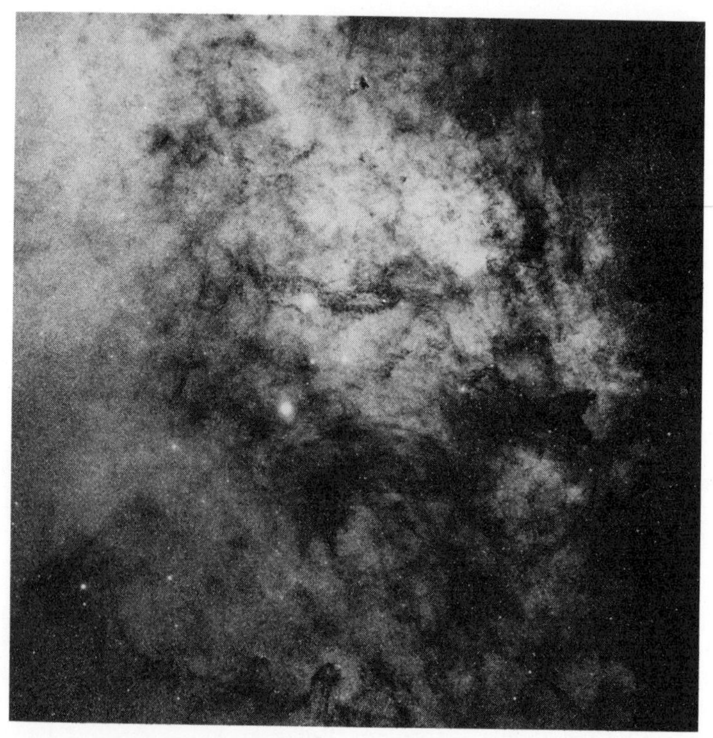

Figure 4. The Milky Way towards the galactic centre. Baade's Window is the dense star cloud, bounded by dark absorption lanes, at the very centre of the photograph. Below and to the left of it is the third magnitude star γ Sagittarii, while just off the picture at upper right (some four degrees away) is the direction of the galactic centre itself. Notice the almost complete absence of stars in this area, indicating very heavy absorption by intervening dust clouds. (Photograph courtesy UK Schmidt Telescope Unit, © Royal Observatory, Edinburgh)

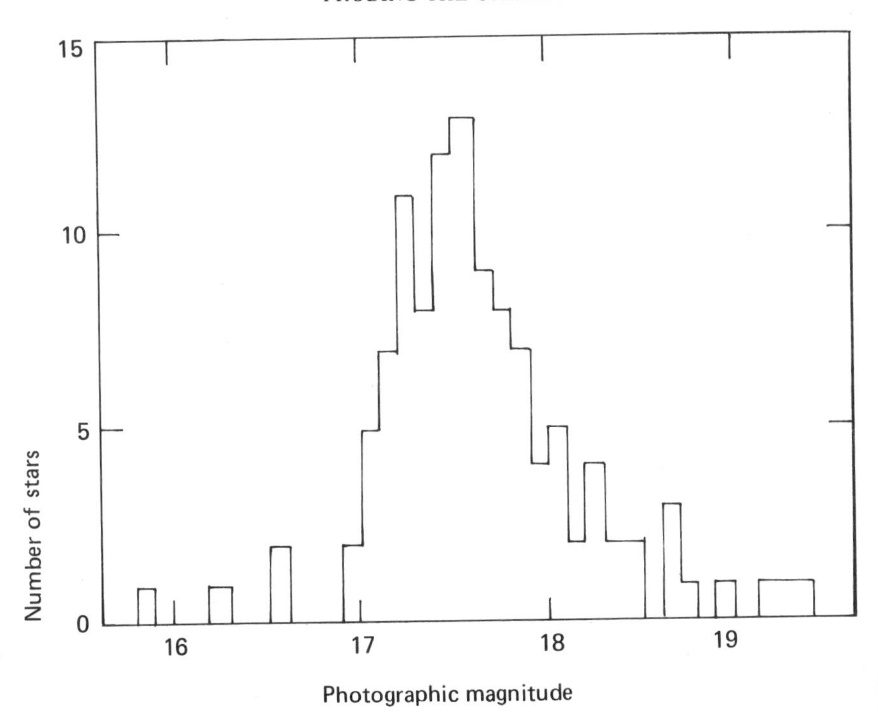

Figure 5. Number–magnitude diagram for the RR Lyræ stars in Baade's Window, showing a peak at about 17m.5 resulting from the high density of these stars near the galactic nucleus. The diagram has been plotted using photographic, or blue magnitude; for RR Lyræs this quantity is about 0m.2 fainter than the visual (yellow) magnitude $<m_V>$.

(Figure 5), he found that there was a peak near 17m.5 which he attributed to the strong concentration of stars around the galactic nucleus. When translated into distance, that peak occurs at 8.1 kiloparsecs (26,000 light-years), which gives us a direct measure of the Sun's distance from the galactic centre. Baade's calculation was based on a value of $<M_V>_{RR}$ of $-0^m.2$; if we adopt the value of $+1^m.0$ which I mentioned earlier, together with modern estimates of the absorption, the galactic centre distance reduces to about seven kiloparsecs.

The most important survey of the galactic bulge carried out to date will probably be one of the last to use purely visual methods in locating the variables on the photographic plates, as distinct from the more efficient automatic means we have available today. It was made in the 1960s and 70s by L. Plaut at Gröningen in the Netherlands using Palomar Schmidt plates, and is hence generally known as the Palomar–Gröningen Survey. Plaut identified three lines of sight (calling them fields 1, 2, and 3) which are 29° above, 12° above, and 10° below the galactic centre, respectively. He found a total of 1183 RR Lyræs in the three fields. Since it is fairly safe to assume that the galactic halo is symmetrical about the plane of the disk (so that things are the same 10° above as they are 10° below), the density distribution of RR Lyræs along each line of sight can be combined to produce a detailed picture of the inner galactic halo and the bulge. The analysis of these data, together with similar data from Baade's Window, was carried out by Plaut in collaboration with Jan Oort of Leiden, and their results appeared in a major paper in 1975.

Some years ago, Victor Clube and I carried out a re-analysis of Plaut's data, using the revised value of $<M_V>_{RR} = +1^m.0$. The picture that emerged when we plotted contours of constant RR

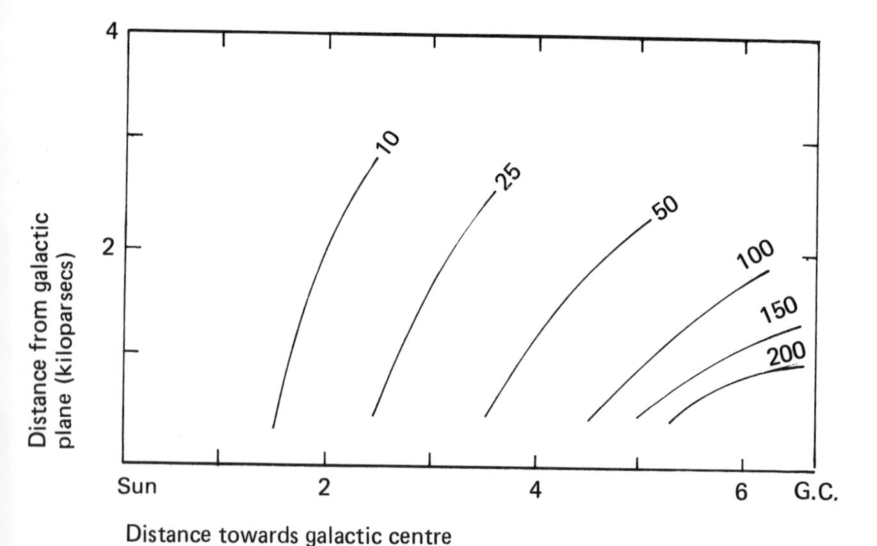

Figure 6. Contours of equal space–density of a- and b-type RR Lyræ stars in a vertical cross-section through the inner galactic halo. The density levels are in stars per cubic kiloparsec (a volume equivalent to 35 billion cubic light-years).

Lyræ density in a slice through the galactic halo is shown in Figure 6. This plot has been 'folded' so that contours on the far side of the galactic centre match up with the ones on this side, a principle of symmetry that allows us to determine accurately the distance to the galactic centre. Once again, this comes out to be something in the region of 7 kiloparsecs. The density contours which we found confirm the spheroidal nature of the galactic halo, and point to a very rapid rise in the RR Lyræ density as one approaches the nucleus. Results like these, together with the more elaborate conclusions of Oort and Plaut, are important to theoretical astronomers building mathematical models of the Galaxy, because they place constraints on the mass and velocity distributions of galactic material, thus enabling unrealistic models to be identified and rejected.

If we can extend our observations of RR Lyræ stars to include spectroscopic as well as photometric data there is more that we can learn about the nature of the galactic halo. However, because of the difficulties involved in obtaining spectra of faint stars, we are generally confined to sampling objects which are relatively bright – magnitudes 8 to 13 approximately – and therefore rather nearby.

When we examine the spectra of these RR Lyræs, we find they have radial velocities (corrected for pulsation) which range from zero up to two or three hundred kilometres per second. Coupled with this is a large variation in metal deficiency, ΔS, and, interestingly, the stars with the highest metal deficiencies seem to have the highest velocities. Now, the velocities turn out to be predominantly in the opposite direction to that of the Sun, which, you will recall, moves around the galactic centre at some 250 km/s, so the implication is that the highest velocity stars are really those which are most nearly stationary with respect to the Galaxy as a whole. These stars are, in fact, moving in highly elongated orbits which carry them far into the halo, and are thus members of the outermost population of our Galaxy. As we have seen, they are also very metal-weak. So, what are they? Well, the accepted interpretation of all this is that these RR Lyræ stars are relics of the very earliest period in our Galaxy's history.

We have seen that RR Lyræ variables are extremely old objects – probably more than 10^{10} years – and it is reasonable to assume that the most ancient will be metal-poor compared with a body like the Sun (a mere 4.6×10^9 years old) because the gas from which they formed had not been enriched with heavy elements to the same extent as that from which the Sun formed. The fact that we see these galactic fossils high in the halo implies that they might have come into existence when our Galaxy was collapsing from a spherical gas cloud towards the flattened system we see today. The added fact that we find low-velocity, low ΔS (and presumably younger) RR Lyræs in the disk adds weight to this hypothesis.

This is the 'standard' view of the origin of the galactic halo and disk, but there are other theories. One, suggested by Unsöld and others, envisages the explosion of exotic supermassive stars (perhaps 10^9 times the mass of the Sun!) in the galactic nucleus. In this scenario, the cores of such objects, having an exceedingly high temperature and low angular momentum (or spin) become the raw material for the halo, while the cooler, higher angular momentum outer layers eventually form objects in the galactic disk. The observed metallicity and velocity differences between the halo and the disk are, in this picture, a direct consequence of temperature and angular momentum differences of material within the supermassive objects. In the light of what I am going to say in the concluding section of this article, it is interesting to note that successive explosive events of this kind in the galactic nucleus would give rise to expanding and contracting motions in the Galaxy.

That, then, is something of the story of RR Lyræ variables and the way in which they have shed light on the geography and history of the Galaxy. Our discussion has dwelt principally on the so-called 'field' RR Lyræs – the RR Lyræs occurring individually in the halo and disk – but no article purporting to be about probing the Galaxy with RR Lyræs would be complete without some mention of their counterparts in the globular clusters: Bailey's original 'cluster variables'. In fact, an account of variable stars in globular clusters would be a long story in itself. The RR Lyræs have been much used as primary distance indicators for the clusters; indeed, it was the RR Lyræs that first enabled Shapley, in 1918, to determine the true three-dimensional shape of the galactic globular cluster system. Our current picture of this – a spheroidal swarm of 150 or so globulars orbiting in the galactic halo – is the direct descendant of Shapley's work. Our knowledge of the chemical composition and masses of stars in globular clusters also owes much to the presence of RR Lyræ variables and, indeed, these parameters could not easily be determined without them.

I should like to end this article with one last example of the way in which we can use observations of RR Lyræs to learn about some of the most fundamental properties of our Galaxy. It is a project with which I am involved at the Royal Observatory, Edinburgh, and which promises ultimately to yield some quite exciting results.

Is the Galaxy expanding?

The conventional picture that we have of the motions of objects in both the disk and the halo of our Galaxy is that they are in a state of rotation about the galactic centre with no overall radial motion – that is, no net expansion or contraction of the system. Thus, while some objects might be in orbits taking them in towards the galactic centre, a roughly equal number will be in outward-bound orbits, and the system is said to be 'relaxed'. This hypothesis has the merit of considerable dynamical simplicity and, indeed, appears to be supported by a number of observational facts.

Some 21 years ago, the radio astronomer F. J. Kerr found that he had a problem in matching the velocities of hydrogen clouds in the northern and southern hemispheres of the sky. The only way he could resolve the difficulty was to postulate an outward motion of that part of the Galaxy in which the Sun is situated, amounting to some 7 km/s. Hitherto, the so-called 'local standard of rest' at the Sun had been assumed to be moving in a perfectly circular orbit, and this was the first time anyone had seriously suggested it might have a radial motion in the Galaxy. But a dozen years later, my colleague, Victor Clube, discovered evidence for a much larger outward motion in a new catalogue of stellar proper motions from the Lick Observatory. His value for the outward velocity of the local standard of rest – usually given the symbol Π_0 – was nearer to 40 km/s, and he soon found that this figure was strongly supported by certain radio observations of hydrogen and molecular features in the galactic centre.

It was more or less at this point that I began working with Clube, and the first thing we discovered was that the galactic globular cluster system – the most obvious component of the halo – is, in its inner regions, displaying expanding motions to the tune of 80 km/s. This effect was later examined by Carlos Frenk and Simon White of Cambridge University in a major study of the globular cluster system; they verified its existence and also noted that the outer regions might be in a state of collapse.

Thus we were – and to a lesser extent still are – presented with two conflicting sets of observations. One says that bulk inward or outward motions of any importance in the Galaxy do not exist and, in particular, that the local standard of rest has a well-behaved circular orbit. The other says that large systematic radial motions are present, and that our bit of the Galaxy is partaking in them with an outward velocity. Which picture is right? Unfortunately, observations of external galaxies do not help us, because about half of them show systematic non-circular motions and the other half do not. There is another question: if the expanding motions in our Galaxy are real, what causes them? Well, I have already hinted at one possible mechanism, but this in turn raises other problems, so that there is no immediate, straightforward answer. There are, however, two steps we can take to decide whether we should believe the evidence for expansion. One involves looking at young stars in the disk, traditionally cited as a class of objects having a purely circular motion, and, indeed, that is currently being done with quite remarkable results – but that is a story for another time. The other is to try and pin down once and for all the radial motion of the local standard of rest – and that can be done with RR Lyræ stars.

The most obvious way of testing whether you are moving inwards or outwards in the Galaxy is to measure your radial velocity with respect to something stationary at the galactic centre. Unfortunately, this is easier said than done, because in the optical region we cannot even see the centre due to heavy obscuration, and radio and infrared observations give results which cannot be described as unambiguous. However, we *can* see objects in the optical which we know are closely connected with the galactic centre, and these are the RR Lyræs in the bulge. If we could secure spectroscopic radial velocities for enough of these, and calculate from the results a mean velocity (taking into account the effects of galactic rotation), then we would have a very good direct estimate of the value of Π_0.

The best candidates for this job are the 100 or so RR Lyræs in Baade's Window but, as we have seen, they are very faint, with magnitudes in the region of $17^m.5$. There are not many telescopes in the world capable of obtaining spectra of such faint stars with the required velocity precision of a few kilometres per second, but fortunately, British astronomers have access to one of them – the 3.9-metre Anglo-Australian Telescope (AAT) at Siding Spring,

New South Wales. This amazing instrument, with its equally amazing Image Photon Counting System developed by Alec Boksenberg at University College, London, makes child's play of measuring these velocities, as long as you can spend thirty minutes to an hour on each star.

Working with Victor Clube, and with invaluable help from Paul Murdin of the Royal Greenwich Observatory, I have obtained spectra of a number of these RR Lyræs over the last three or four years. That number is rather small because, to our intense embarrassment, we have turned out to be the world's unluckiest observers on the AAT! Out of thirteen nights – a fairly generous allocation if a rather ominous number – we have had eleven cloudy ones, with conditions ranging from a thin but completely opaque cover of mountain mist to thick black murk with unprecedented torrential downpours. Let me hasten to add that Siding Spring boasts, on average, 65 per cent of its nights clear, so we represent an anomaly which appears to defy statistics. We are, however, very popular with local farmers, whose water shortage problems invariably vanish with our arrival.

The RR Lyræs for which we have obtained velocities are, as yet, too few to say anything significant about the value of Π_0. However, one tantalizing piece of evidence has emerged because we have also observed some brighter RR Lyræs in one of the Palomar–Gröningen fields. These lie between the Sun and the galactic centre and, without exception, they have negative velocities (i.e., they are moving towards the Sun); they are thus mimicking the rapid expansion of the inner globular cluster system mentioned earlier.

What is needed to complete this programme (apart from some decent weather) is a means of speeding up the rate at which the RR Lyræs can be observed, and a new instrument has recently been developed at the AAT which could do just that. It is simply a fibre-optics coupler, allowing astronomers to obtain spectra of up to 50 objects at the same time. It works because the detector principally used on the AAT, the Image Photon Counting System (IPCS), is rather like an electronic photographic plate in that it records an image over a two-dimensional area. If that image is the spectrum of a single star, then much of the detector is clearly being wasted. The new system will bring in light from many stars in the telescope's field of view by means of optical fibres with their output ends lined up on the spectrograph slit, so that a whole array

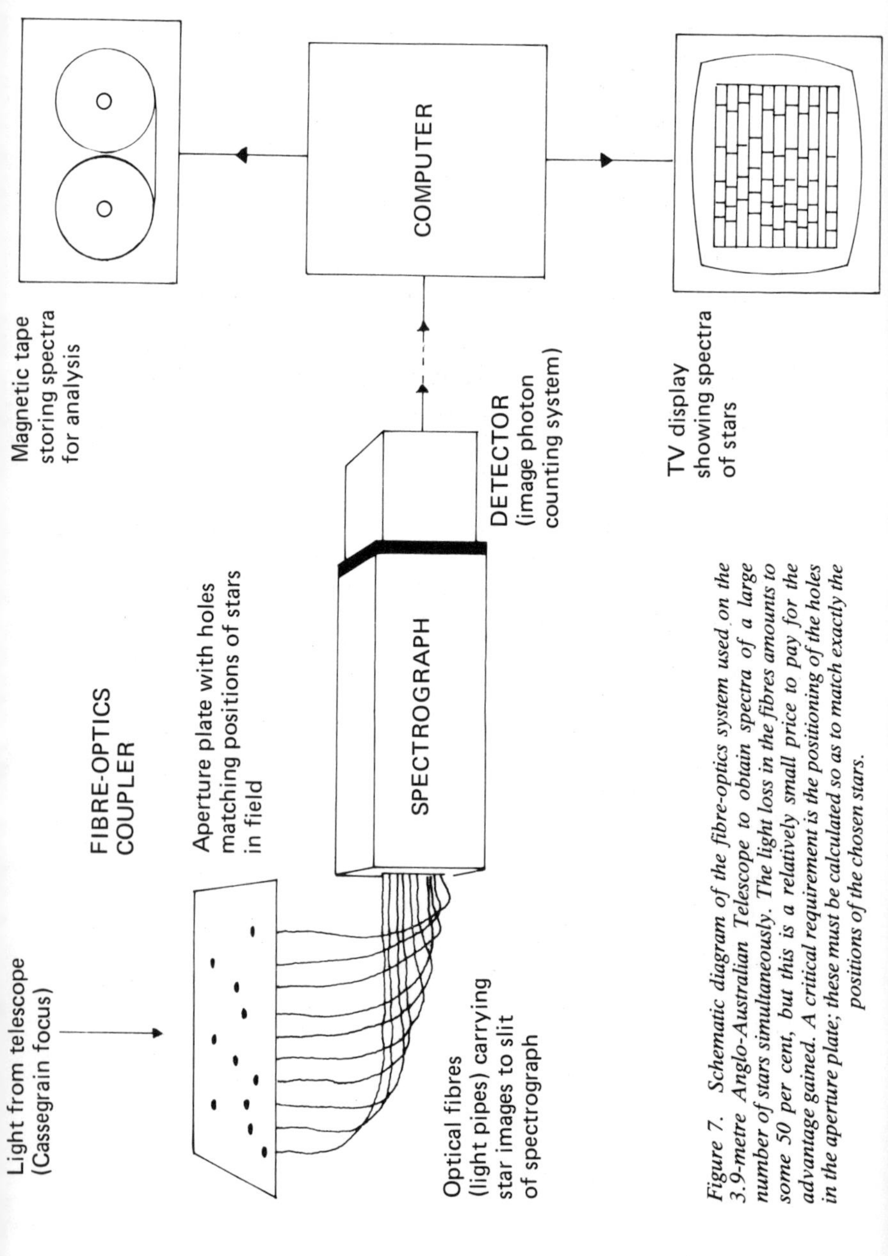

Figure 7. Schematic diagram of the fibre-optics system used on the 3.9-metre Anglo-Australian Telescope to obtain spectra of a large number of stars simultaneously. The light loss in the fibres amounts to some 50 per cent, but this is a relatively small price to pay for the advantage gained. A critical requirement is the positioning of the holes in the aperture plate; these must be calculated so as to match exactly the positions of the chosen stars.

of spectra will fall upon the detector simultaneously (see Figure 7). Of course, one has to remember which spectrum belongs to which star, but that is a minor problem. The big problem is concerned with positioning the input ends of the fibres, because these have to be lined up on the selected stars with an accuracy of about one-third of an arc second. This is achieved by means of an aperture plate which has been very precisely perforated at the calculated positions of the stars; the fibres then lead away from the holes at the back.

The compact nature of Baade's Window makes it an ideal area of the sky in which to use the coupler, and the RR Lyræs are sufficiently dense that we can contemplate obtaining spectra of 20 at a time. It might very well be that this is the breakthrough needed to gather enough data to firmly establish the value of Π_0 and, hopefully, resolve the question of galactic expansion. As an added bonus, each RR Lyræ spectrum will yield a value of $\triangle S$ and give us more information on the chemical composition of the galactic nucleus.

As I write these words, I am looking forward with considerable anticipation to a time about three months hence when I shall be pointing the AAT once again towards Baade's Window. The fibre coupler will be on the telescope and, if it is clear and I have done my sums right, the faint spectra of 20 or so RR Lyræs will be falling on the IPCS photocathode. The results from that observing run might be quite interesting . . .

Watch this space!

Acknowledgements

I should like to thank friends and colleagues at the Royal Observatory, Edinburgh, for their helpful comments on this article. I owe a particular debt of gratitude to Dr Victor Clube, who not only contributed much to the article, but who set me on the trail of RR Lyræ stars in the first place.

The Night Sky – A.D. 50,000

S. A. BELL

This article explores the future of the constellation patterns and the stars that make up those patterns. Tables of the brightest stars for 1,000,000 B.C. and A.D. 1,000,000 are also compiled.

One of the most useful items for the beginner in astronomy is a good star chart. This will help him to recognize the major constellation patterns and the stars which make up those patterns.

The chart he sees before him will look very much the same as Ptolemy saw it when he drew up his original forty-eight patterns nearly two thousand years ago. Since then these patterns have been regarded as 'fixed' because the motions of the stars in most cases are so small that they can only be detected over a period of about a century. However, over longer periods of time, these motions will change the patterns of the night sky to such an extent that certain constellations will fade below naked eye visibility whilst others will be scattered over vast areas of the sky.

How will the night sky of the future differ from that of the present? For instance, what will the night sky look like in 50,000 years hence and how will time affect the patterns we are familiar with today? To take this question to an extreme, which stars were the first magnitude objects in the skies of a million years ago and which ones will become the bright stars of a million years hence.

About a year ago, my University Astronomical Society required a simple star map to help members get acquainted with the night sky. I decided that the quickest and simplest method to do this was to write a computer programme to calculate the positions of a selected number of stars. This selection was made up of those stars involved in the constellation patterns visible from St Andrews at different times during the year. The map had to show not only the position of the stars in azimuth and altitude but also their magnitude as well as their place in a particular constellation pattern.

Approximately 400 stars were involved in these patterns and

plotting such a number by hand would have been extremely tedious. Consequently a completely computerized method of creating and plotting the star map was the most desirable answer.

Simple formulæ are readily available to calculate the azimuth and altitude of a star. These formulæ require the position of a star in Right Ascension and Declination, the latitude of the site from which the observations are being made and the local sidereal time.

An algorithm to provide the local sidereal time was found in *Practical Astronomy with your Calculator*[1], which proved to be very simple to use. Knowing the time and date, Greenwich sidereal time could be calculated which gives the local sidereal time by subtracting the longitude of the site measured positively westwards. It is worth pointing out that this programme uses readily available formulæ and data with the emphasis on the simplicity with which it can be used.

From the results obtained using this initial attempt, it became clear that the method could be extended to all 88 constellations and star maps could be created for almost anywhere in the world. I say 'almost' because the formulæ used to do this break down for the north and south poles. To display all 88 constellations 806 stars are needed.

The positions used for this programme were those of the epoch 1950.0 and it was obvious that to make the map a little more accurate positions would be needed that were corrected for precession. It was at this point that the idea of seeing how the constellations altered over long periods of time occurred to me. This would require knowledge of the movements of the stars which eventually cause the distortion and break up of the patterns we see today.

The Equatorial system of co-ordinates, namely Right Ascension and Declination, is not fixed in space and moves relative to a true inertial frame of reference owing to the effects of precession. This change in the co-ordinate system is caused by the combined gravitational effects of the Sun and the Moon on the non-spherical figure of the Earth. Hence the celestial pole describes a circle of radius 23 degrees 27 minutes around the pole of the ecliptic in a period of just under 26,000 years. The resulting slow westward movement of the vernal equinox brings about a change in the Right Ascension and Declination for every object in the sky.

Simple formulæ to correct for precession tend to break down round the poles, and it was for this reason that rigorous precession formulæ devised by Newcomb at the end of the last century were used in the updated programme. It must be said that these formulæ are only really applicable for a few centuries and were never intended for protracted use.

On a longer time-scale, the cumulative effects of the individual stellar motions become more noticeable. Some stars are approaching the Sun and will become brighter, while others are receding and will fade. Each star in the Galaxy has its own motion reflected by its proper motion and radial velocity. Nearby stars, such as Alpha Centauri and Arcturus, have large proper motions simply because of their proximity to the Sun. Some stars, however, belong to moving groups which started life at a certain time and place in the Galaxy. A good example of this are five members of Ursa Major, namely Beta, Gamma, Delta, Epsilon, and Zeta, which maintain the well-known shape of this constellation for some considerable time. To obtain the position of a star in the future one must have knowledge of its proper motions in Right Ascension and Declination, its distance and also its radial velocity. It is necessary to assume that the star moves in a straight line through space. Using a change of co-ordinate system this calculation can be done quite easily. This method can be used for up to about a million years into the future as well as the past. Since the Galaxy rotates once in about 250 million years, the approximation that a star moves in a straight line in the period suggested is a reasonable one. The limit is set by the effects of galactic rotation but this complication does not significantly change the star's position during the two million year period in question. It is therefore ignored in this programme.

The source of the data used for this version of the programme was the *Sky Catalogue 2000.0*[2] which contains the necessary data for the epoch 2000.0. Where gaps existed in the information required, e.g., radial velocities and distances, reference was made to more specialized catalogues. Combined magnitudes have been quoted for double stars that occur in the constellation patterns, and where variable stars occur these magnitudes have been quoted for maximum light. It has also been assumed that stars remain constant in output over the period considered, and in the case of the variable stars their maxima have also been assumed constant.

Having incorporated these routines into the programme, precession could be used for up to ten thousand years into the future or the past and what could be called 'space motions' used for the rest of the time. Over substantial periods of time precession is neglected in favour of the more noticeable effects of the intrinsic stellar motions. The programme could now give reasonably accurate positions for the stars at any time between A.D. 1,000,000 and 1,000,000 B.C.

The map produced is a 180-degree × 90-degree rectangular plot which gives a reasonable representation of the night sky up to an altitude of about 60 degrees. Above this limit constellations are liable to be stretched depending on their shape. More sophisticated map projections could no doubt be used but limitations in the plotting system currently available for this programme preclude their use.

The patterns used to delineate the constellations follow those used in *Sky Atlas 2000.0*[3]. There are many different ways to join up the stars of the night sky and those used by this atlas seem to depict the figures they are supposed to represent quite well. The magnitudes of the stars are simply plotted as dots, the larger the dot the brighter the star.

In spite of the inherent difficulties in the map projection being used here, every reasonable attempt has been made to make the star map appear as realistic as possible. One such refinement is the correction for atmospheric extinction. This effect is caused by the scattering of star light by the Earth's atmosphere which is most noticeable when a bright star is near the horizon. It appears considerably dimmer than it would if it was high in the sky. This extinction can be represented quite well by trigonometrical functions of the altitude of the object being observed, although the relation begins to break down at altitudes of less than 10 degrees. For this range of altitudes extinction tables are available for most observatories in the world and a representative average has been used for a sea level site in this programme.

The four star maps in this article have been produced to show the sky as it looks at the moment, and as it will appear in A.D. 50,000. The maps shown have been produced for Central England (latitude 52 degrees north and longitude 0 degrees) and Eastern Australia (latitude 35 degrees south and longitude 150 degrees east). Two of them use the precession routine, namely one for Christmas Day 1984 in England and the other for the 1st July 1984 in Australia. The other two use the 'space motions' routine and are set up for 1st July A.D. 50,000 in Australia, and 1st January A.D. 50,000 in England. All the maps are chosen to appear as the sky does at 22:30 local time after having been corrected for atmospheric extinction. The direction in which they point is given at the bottom left hand corner.

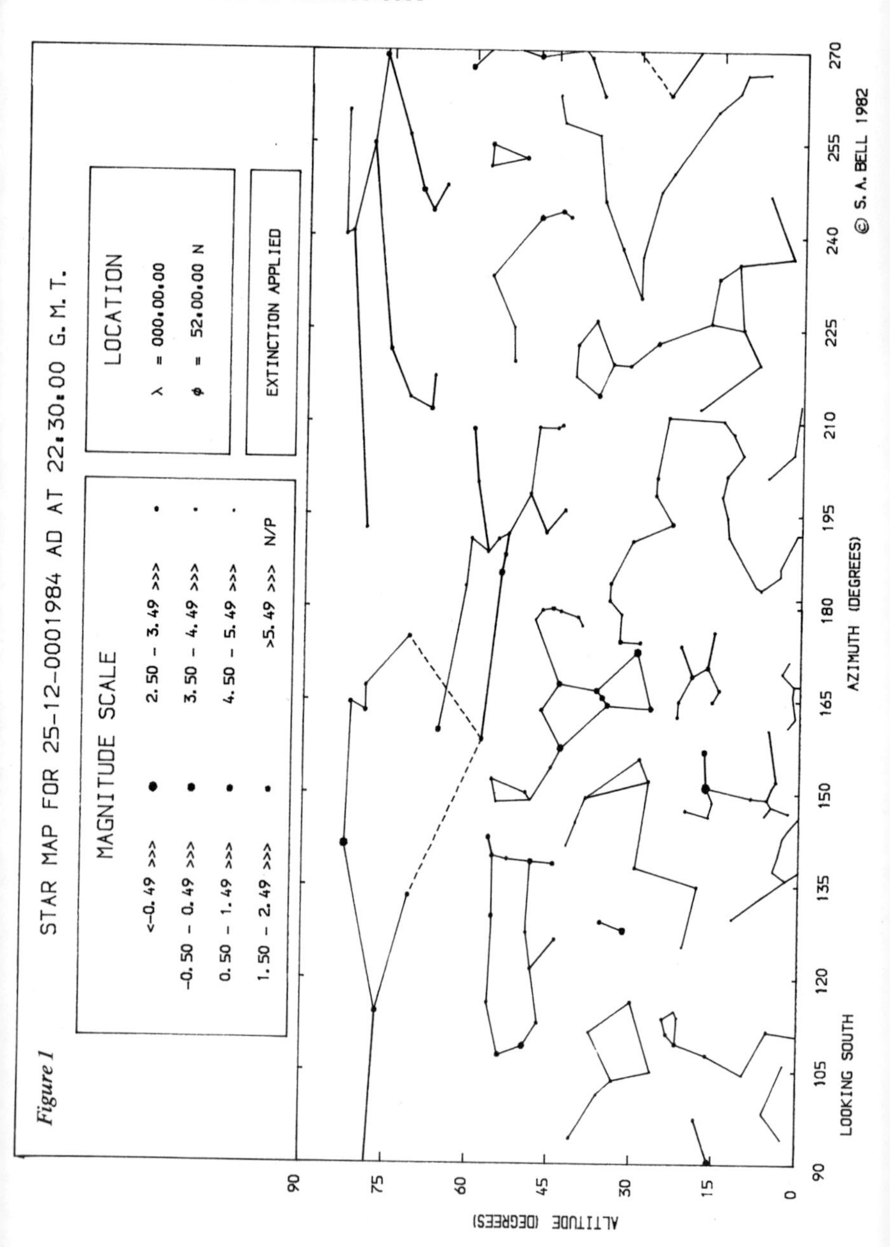
STAR MAP FOR 25-12-0001984 AD AT 22:30:00 G. M. T.
MAGNITUDE SCALE
<-0.49 >>>
-0.50 - 0.49 >>>
0.50 - 1.49 >>>
1.50 - 2.49 >>>
2.50 - 3.49 >>>
3.50 - 4.49 >>>
4.50 - 5.49 >>>
>5.49 >>> N/P
LOCATION
λ = 000.00.00
φ = 52.00.00 N
EXTINCTION APPLIED
ALTITUDE (DEGREES)
90
75
60
45
30
15
0
90
105
120
135
150
165
180
195
210
225
240
255
270
LOOKING SOUTH
AZIMUTH (DEGREES)
© S. A. BELL 1982

Figure 1

Figure 2

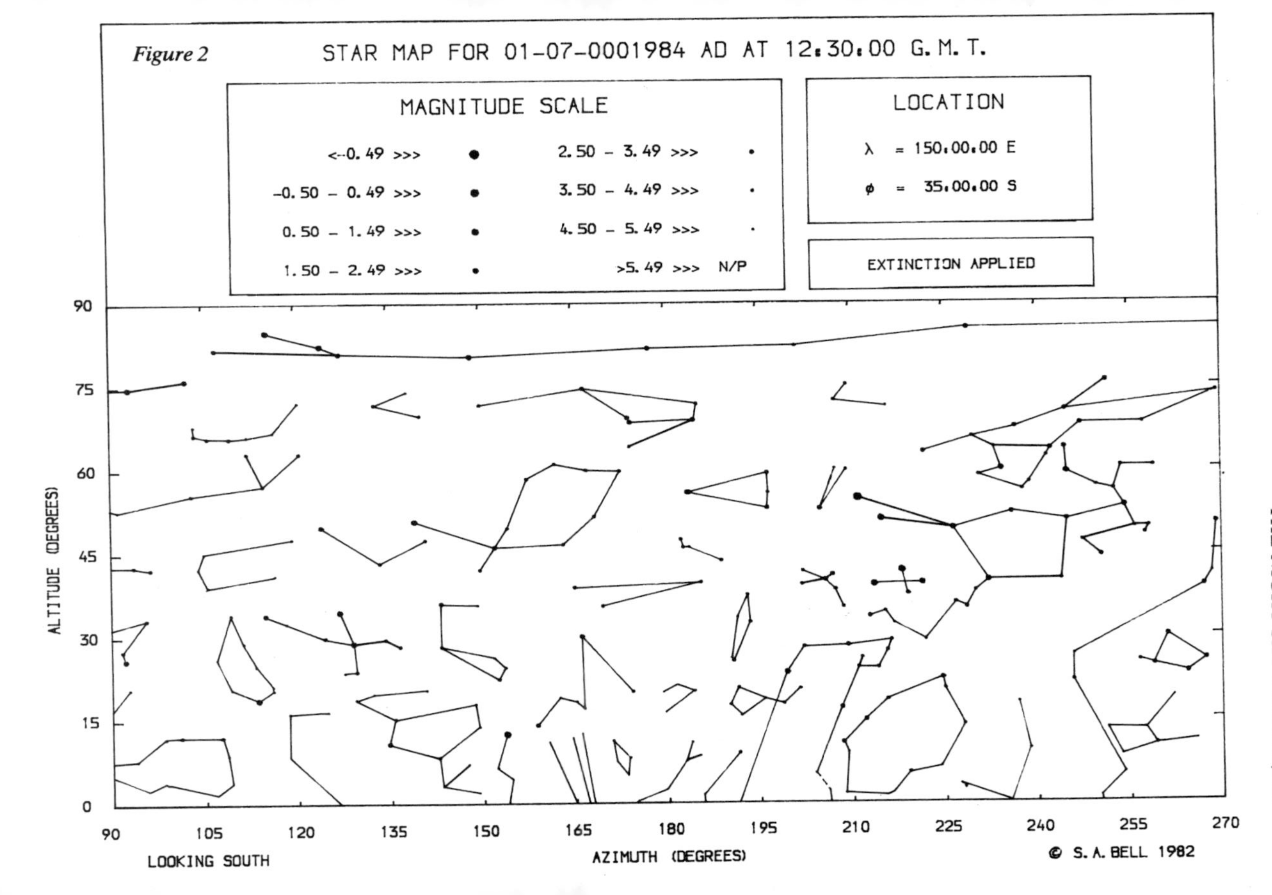
STAR MAP FOR 01-07-0001984 AD AT 12:30:00 G.M.T.
MAGNITUDE SCALE
<-0.49 >>>
-0.50 - 0.49 >>>
0.50 - 1.49 >>>
1.50 - 2.49 >>>
2.50 - 3.49 >>>
3.50 - 4.49 >>>
4.50 - 5.49 >>>
>5.49 >>> N/P
LOCATION
λ = 150.00.00 E
φ = 35.00.00 S
EXTINCTION APPLIED
90 75 60 45 30 15 0
ALTITUDE (DEGREES)
90 105 120 135 150 165 180 195 210 225 240 255 270
LOOKING SOUTH
AZIMUTH (DEGREES)
© S. A. BELL 1982

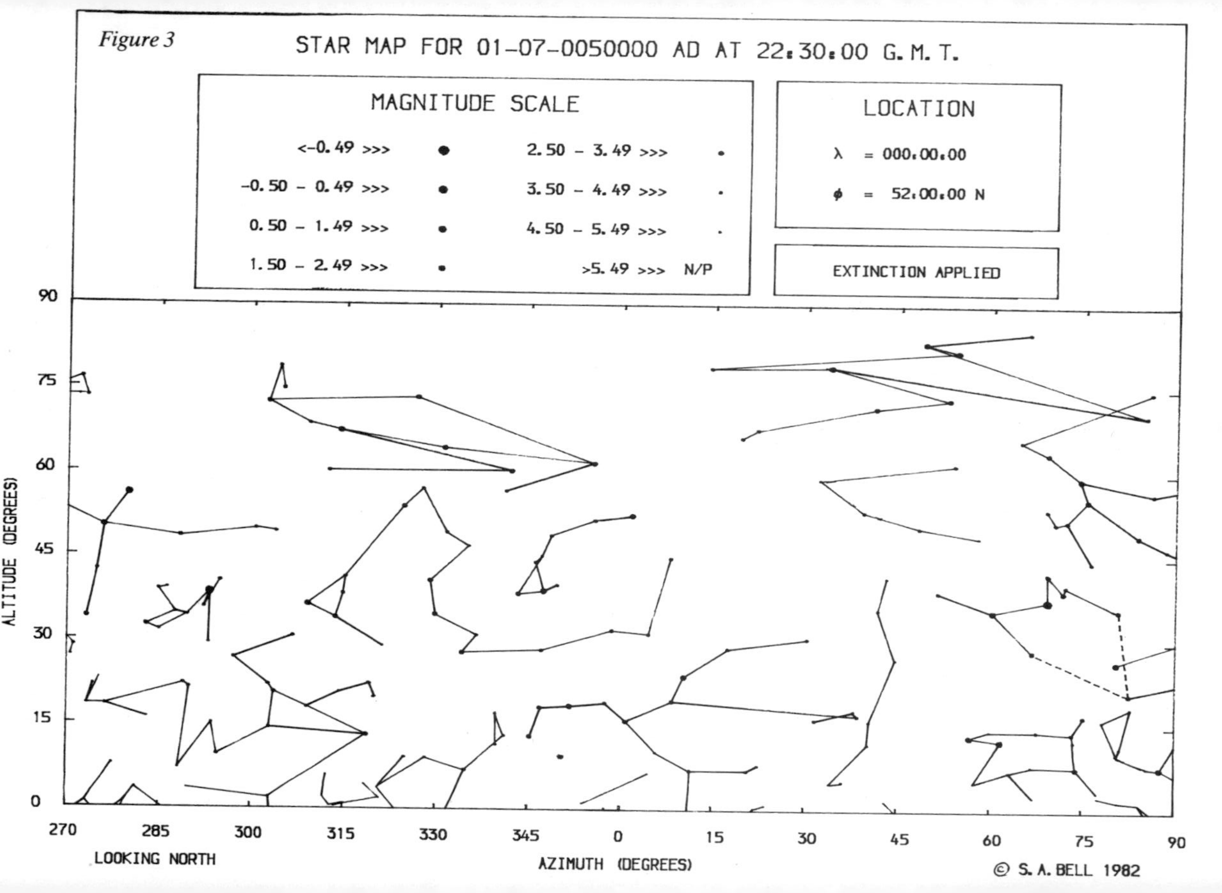
STAR MAP FOR 01-07-0050000 AD AT 22:30:00 G. M. T.
MAGNITUDE SCALE
<-0.49 >>>
-0.50 - 0.49 >>>
0.50 - 1.49 >>>
1.50 - 2.49 >>>
2.50 - 3.49 >>>
3.50 - 4.49 >>>
4.50 - 5.49 >>>
>5.49 >>> N/P
LOCATION
λ = 000:00:00
ϕ = 52:00:00 N
EXTINCTION APPLIED
ALTITUDE (DEGREES)
0
15
30
45
60
75
90
270
285
300
315
330
345
0
15
30
45
60
75
90
AZIMUTH (DEGREES)
LOOKING NORTH
© S. A. BELL 1982

Figure 3

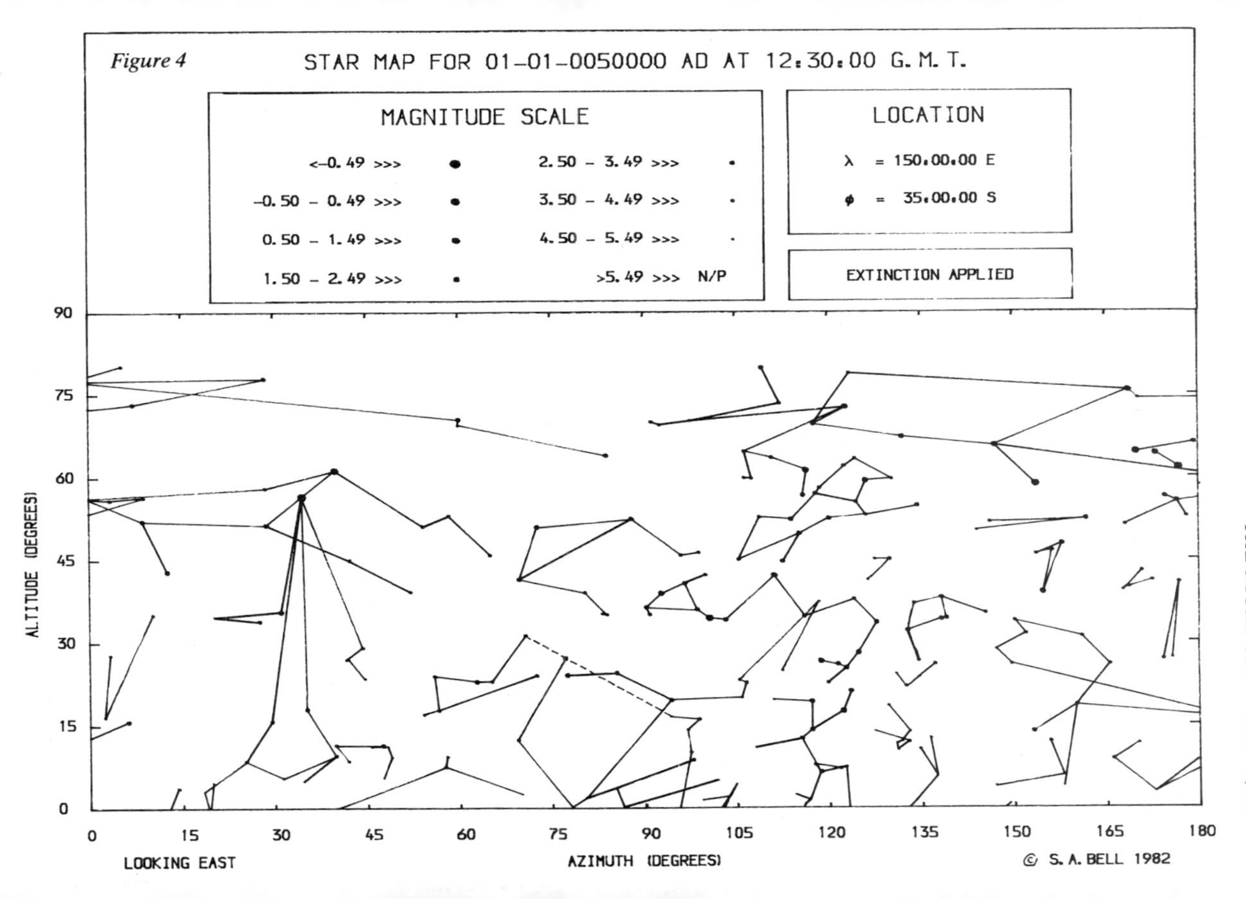
STAR MAP FOR 01-01-0050000 AD AT 12:30:00 G.M.T.
MAGNITUDE SCALE
<-0.49 >>>
-0.50 - 0.49 >>>
0.50 - 1.49 >>>
1.50 - 2.49 >>>
2.50 - 3.49 >>>
3.50 - 4.49 >>>
4.50 - 5.49 >>>
>5.49 >>> N/P
LOCATION
λ = 150:00:00 E
φ = 35:00:00 S
EXTINCTION APPLIED
ALTITUDE (DEGREES)
90
75
60
45
30
15
0
0
15
30
45
60
75
90
105
120
135
150
165
180
LOOKING EAST
AZIMUTH (DEGREES)
© S. A. BELL 1982

Figure 4

Having outlined the way in which the star maps are created, it is time to go back to the questions asked at the beginning of the article. I thought it may be of interest to find out what happens to the first magnitude stars of today. Some of these stars are a substantial distance away from the Sun and little change can really be expected. It is among the nearby stars that changes will become most noticeable. The reference to a star being in a given constellation is based on the boundaries that exist at the present.

The brightest star in the night sky is Sirius, but this has not always been the case. A million years ago it was a third magnitude star in Lynx, and will reach its maximum brightness of -1.7 in just over 60,000 years' time. In the years after that it begins to fade, and will eventually become a third magnitude star once again in the southern constellation of Indus in a million years' time. Its nearest rival, Canopus, remains a first magnitude star over the two million year period in consideration. Alpha Centauri has a very different story. It became visible to the naked eye 900,000 years ago, and will fade beyond that limit again in another 900,000 years' time. It will be seen to its best advantage in nearly 30,000 years' time when it reaches -1.0. It also moves through a large portion of the sky during that time, starting in the southern constellation of Telescopium and ending up in Auriga.

Vega, one of the 'summer triangle' stars for northern observers, is gradually brightening and reaches its best in 300,000 years' time while in the constellation of Cassiopeia. On the other hand Capella is now past its best, and will become a very ordinary star in Orion in a million years' time. Both first magnitude members of Orion fade very slightly, but still retain their first magnitude status.

Procyon in Canis Minor reaches its best in a mere 10,000 years' time, but even then does not really brighten a great deal. Like Capella, it fades to around the third magnitude in a million years' time. Achernar, the only really significant star in Eridanus, fades to nearly second magnitude by the end of the period examined here, making this constellation a rather dull, meandering line of stars.

Some stars escape the ravages of time to a large extent and remain more or less where they are today. Both Acrux and Mimosa in Crux Australis and Beta Centauri brighten over the two-million-year period by a small amount, and this trend is also shared by Deneb and Antares. Spica, on the other hand, shows no real change in brightness.

Altair reaches −0.6 in another 150,000 years' time, when it will lie in Andromeda. As for Regulus and Pollux, they are past their best and are heading towards life as second and third magnitude stars in Monoceros and Cetus respectively. Fomalhaut in Pisces Austrinus was at its best nearly a quarter of a million years ago, when it reached a magnitude of 0.85 in Capricornus.

The last of the recognized first magnitude stars of the present is Aldebaran, which was probably the brightest star in the skies above our Stone-age ancestors when it was in the constellation of Cepheus.

In many books lists can be found of the brightest stars in the night sky. The list for A.D. 50,000 contains the same stars that are found in the compilations for the present. It should be pointed out that some stars in the middle and at the end of that list do change position although the brightest three retain their current positions. If these listings are compiled for both 1,000,000 B.C. and A.D. 1,000,000 then new entries can be expected. Tables I and II show this quite clearly. A rough idea of where the star is in the sky is given by the name in the constellation column.

A point that must be made here when dealing with these compilations for great distances into the past or the future is that there are just over 8,000 naked-eye stars in the night sky at present. The sample used for this programme is 10 per cent of that number, so there is a possibility that a star, or indeed stars, in the remaining 90 per cent may become brighter than those listed in the tables. It is less likely that a star below naked-eye visibility will make a long lasting impact on the night sky since intrinsically bright nearby stars are something of a rarity.

The stars which appear in both the tables are most likely to be the more distant bright stars. Good examples of this are Canopus, Rigel, and Betelgeux, which are all very slowly receding from us.

Finally, the answer to the question of how the constellation patterns change over the next 50,000 years. The types of change can be split into three categories. The first category covers those patterns which are only minimally affected by time, the second covers those which have one or two stars which distort the pattern or change noticeably in magnitude and the third category deals with a less frequent occurrence where a constellation is so distorted that it becomes unrecognizable. Obviously the third category becomes the most popular over really long periods of time, but over the quoted period the second category is likely to be the most

TABLE I

The Twenty Brightest Stars for A.D.1,000,000

	Star name	*Magnitude*	*Constellation*
1	Gamma Draconis	−2.18	Ophiuchus
2	Delta Scuti	−1.32	Scutum
3	Beta Aurigæ	−0.68	Pisces
4	Gamma Scuti	−0.67	Telescopium
5	Alpha Carinæ	−0.56	Puppis
6	Delta Sagittarii	0.08	Grus
7	Beta Orionis	0.29	Orion
8	Beta Centauri	0.45	Musca
9	Gamma Geminorum	0.69	Virgo
10=	1 Centauri	0.79	Centaurus
10=	Alpha Crucis	0.79	Carina
12	Alpha Scorpii	0.90	Scorpius
13	Alpha Ceti	0.94	Fornax/Phœnix
14	Alpha Orionis	0.97	Orion
15	Alpha Virginis	1.08	Corvus
16	Beta Ursæ Majoris	1.09	Hercules
17	Epsilon Ursæ Majoris	1.11	Hercules
18	Beta Libræ	1.15	Hydra
19	Epsilon Herculis	1.21	Leo/Leo Minor
20	Alpha Cygni	1.23	Cygnus

TABLE II

The Twenty Brightest Stars for 1,000,000 B.C.

	Star name	*Magnitude*	*Constellation*
1	Kappa Orionis	−4.27	Draco/Ursa Minor
2	Zeta Sagittarii	−2.71	Aquarius
3	Alpha Columbæ	−1.51	Auriga
4	Alpha Carinæ	−0.83	Dorado/Pictor
5	Alpha Eridani	−0.69	Aquarius/Capricornus
6	Gamma Velorum	−0.25	Vela
7	Beta Orionis	−0.06	Orion
8	Alpha Orionis	−0.03	Orion
9	Zeta Leporis	0.08	Hydra
10	Beta Ursæ Majoris	0.29	Ursa Major
11	Zeta Volantis	0.45	Sculptor
12	Beta Centauri	0.80	Circinus/Lupus
13	Alpha Tucanæ	0.95	Perseus
14	Alpha Virginis	1.02	Virgo
15	Alpha Crucis	1.03	Centaurus
16=	Epsilon Canis Majoris	1.04	Canis Majoris
16=	Alpha Scorpii	1.04	Ophiuchus/Scorpius
18	Theta Eridani	1.06	Columba/Puppis
19	Theta Aurigæ	1.07	Vulpecula
20	Gamma Orionis	1.24	Orion

popular. Constellations that have not been mentioned here fit the first category.

Moving alphabetically through the constellations, the first to suffer the effects of time is Aquila. Due to its motion northwards, Altair loses its two fainter companions and becomes half a magnitude brighter than it is at present. On the other hand, Beta moves southwards which gives rise to a 20-degree separation between the two leading stars.

Auriga, the Charioteer, in general moves bodily south in the sky by a small amount, and the kite-like shape gets a little crushed by the more rapid movement of its northerly members, notably Capella. The Hædi or 'The Kids' slowly start to split up caused mainly by the movement of Epsilon. On the other hand Boötes undergoes a north–south stretch caused mainly by Arcturus moving south. At this time Arcturus will be less than five degrees away from Spica, which will provide a striking contrast between these two first magnitude stars because of the orange colour of Arcturus and the blue white of Spica.

The faint but discernible pattern of Capricornus, which resembles an inverted triangle, will be disrupted by the movement of Delta, its brightest star. It moves south by about 12 degrees. Like Capricornus, Cepheus is not a very prominent constellation at the moment, but will improve over the next 50,000 years. Eta brightens by nearly a magnitude, while Gamma brightens by half a magnitude.

Alpha Centauri has already shown itself to be a rapidly moving object, and at this time it resides in the head of Hydra at magnitude −0.6. By A.D. 30,000 this star will become visible in Britain. Cetus, the Whale, now resembles a kite and its tail, since the body becomes a line of stars while the head remains more or less intact. Northern observers, having gained Alpha Centauri, have to lose a first magnitude star in return. Sirius will no longer be visible in Scotland, and barely struggles above the horizon in Southern England. This is unfortunate, since it will then nearly be at its brightest.

The smaller companion to Canis Major, Canis Minor becomes unrecognizable since the two stars are more than 20 degrees apart. Corona Australis, the Southern Crown, loses its shape and resembles a figure eight more than anything else. The most famous small constellation in the sky, Crux Australis, no longer resembles a cross or a kite depending on your interpretation. It should be renamed Corona Australis, because it resembles a bright curved line of stars.

The small but conspicuous constellation of Delphinus takes the shape of a triangle caused mainly by the movement of Delta. Hercules, however, becomes an unrecognizable jumble of stars. Whatever shape there is today completely disappears. It is best described as an area of moderately bright stars. Lepus the Hare, looks very much as though Orion has put his foot on it and crushed it. Moving northwards is Lyra, lead by its brightest star Vega, which is straying away from the faint but discernible rectangle of stars nearby. Returning to the southern hemisphere, Pavo, one of the most conspicuous of the southern birds loses Delta to the nearby constellation of Tucana, making Pavo appear more elongated.

Both the head and the tail of Serpens get distorted, and the head loses one of its brighter members to Libra. Moving southwards, Triangulum Australe becomes a poor geometrical description, since it is better described as a square.

The last constellation in this round-up is Ursa Major, one of the most useful sign-posts in the night sky to anyone trying to find his way around the northern-hemisphere constellations. Unfortunately it will no longer be able to perform this function, partly because other constellations have moved and partly because Alpha and Eta are slowly moving away from the main group. I say 'group' because the other five stars making up the familiar shape are moving through space as a group.

That concludes this glimpse into the skies of A.D. 50,000. No doubt the skies of the far future will look very different to this and at some time a new identification system will have to be brought in. Maybe someone will create new constellations although I very much doubt they will have the colourful imagination of our ancestors who first created and named the patterns we have today.

Star Maps created for this Article

Figure 1. Map created for 25/12/1984 at 22:30 G.M.T. looking south. Location 52 degrees north on the Greenwich Meridian. Extinction applied and patterns drawn.

Figure 2. Map created for 01/07/1984 at 22:30 L.M.T. looking south. Location 35 degrees south at longitude 150 degrees east. Extinction applied and patterns drawn.

Figure 3. Map created for 01/07/50000 at 22:30 G.M.T. looking north. Location 52 degrees north on the Greenwich Meridian. Extinction applied and patterns drawn.

Figure 4. Map created for 01/01/50000 at 22:30 L.M.T. looking east. Location 35 degrees south at longitude 150 degrees east. Extinction applied and patterns drawn.

References

1. *Practical Astronomy with your Calculator*, 2nd Edition, Peter Duffett-Smith, Cambridge University Press.
2. *Sky Catalogue 2000.0*. Edited by Alan Hirchfeld and Roger Sinnott, Cambridge University Press & Sky Publishing Corporation.
3. *Sky Atlas 2000.0*. Wil Tirion, Cambridge University Press & Sky Publishing Corporation.

Edmond Halley: The Man

PATRICK MOORE

As I write these words (March 1983) Halley's Comet, the most famous of these strange cosmic wanderers, has been recovered. It is on its way back to the Sun for the first time since 1910, and although it is still excessively faint it will brighten steadily, finally reaching naked-eye visibility about November 1985. Perihelion is due in February 1986. From a purely spectacular point of view, this is not a good return; the Earth and the comet are in the wrong places at the wrong times. But scientifically the return is of tremendous importance. Sadly, the Americans have abandoned their proposed comet probe on the grounds of expense (surely a most regrettable decision), but four probes are due to be sent to the comet: Europe's Giotto, Japan's Planet A, and the two Russian Vega vehicles, which will go to Halley's Comet by way of Venus. If any of these probes can succeed, we may have our first view of a cometary nucleus.

Meanwhile, let us take a look at the man himself – one of the most eminent of all British astronomers, and who would be remembered more generally than he actually is but for the fact that he lived at the same time as Isaac Newton.

Edmond Halley (not Edm*u*nd, please note) was born on 29 October 1656 at Shoreditch, in London. There were no money problems; Edmond was sent to St Paul's School, becoming captain of the school in 1671, and then entered Queen's College, Oxford. His ability was not in doubt, and his pleasant personality made him popular everywhere. He became interested in astronomy at an early stage in his career, and actually left Oxford in 1676, before completing his degree, in order to undertake a survey of the southern stars, which had been sadly neglected. Accordingly, Halley went to St Helena and carried out systematic observations – despite an unfavourable climate and some decidedly un-cooperative officials. He returned home in 1678, and presented his catalogue of the southern stars to the Royal Society, which august body promptly elected him a Fellow.

His reputation was made, and he began a long series of valuable contributions. He was sent to Danzig (the modern Gdańsk) in Poland to see the famous astronomer Hevelius, and about the same time he proposed a method of using transits of Venus to measure the length of the astronomical unit, or Earth–Sun distance. He was a practical observer and, in particular, he observed the bright comets of 1680 and 1682. It was the latter comet which now bears his name, though it was not until years later that Halley turned his attention to computing its orbit.

By now he had become friendly with Isaac Newton. This was fortunate indeed in view of what happened later. There had been discussions between Halley and his colleagues, notably Robert Hooke and Christopher Wren, about problems of gravitation; there were some calculations which were, for the time, excessively difficult, and Halley realized that only Newton was capable of making them. When Halley made a journey to Cambridge to discuss the problems, he found that Newton had actually solved them some time earlier, but had not published his results and had even mislaid his calculations! At Halley's insistence, Newton reworked the calculations, and the result was the book always known as the *Principia*, sometimes described as the greatest mental effort ever made by one man. The book was published at Halley's expense. Without him, it might never have been produced.

Unfortunately his relations with another famous astronomer. John Flamsteed, were less happy. Flamsteed was a capable observer, and had been installed at the newly founded Royal Observatory, Greenwich, specifically to compile an accurate catalogue of the stars for use by British seamen. Flamsteed was a perfectionist; he refused to publish his work until he was completely satisfied with it, and other astronomers became tired of waiting. Finally, Flamsteed handed the Royal Society committee a copy of his observations, as well as an incomplete manuscript of the catalogue. He made it clear that the catalogue was not to be printed as it stood; the observations, however, could be published.

Time passed, and still Flamsteed did not submit his finished catalogue. Eventually, in 1711, the Royal Society published not only the observations (which Flamsteed had agreed) but also the catalogue (which he had not). To make matters worse, the catalogue carried a preface by Halley which Flamsteed regarded as harmful to his reputation. He went so far as to obtain as many copies of the catalogue as he could, and burn them publicly to show 'the ingratitude' of two of his countrymen – Newton and Halley. The quarrel was never patched up; Flamsteed died in 1719, and the final version of the catalogue was published posthumously seven years later.

Flamsteed had been the first Astronomer Royal. On his death Halley was the obvious successor, and at once he began a long series of observations of the motions of the Moon which were essential for the navigational method which had been worked out for British seamen. He had various problems to face; for instance he had to re-equip Greenwich Observatory, since the original instruments had been provided by Flamsteed, and Mrs Flamsteed descended upon the observatory like an east wind, claiming the instruments and removing them (with the sad result that they have been permanently lost). Halley, however was equal to the task, and had the satisfaction of completing the entire series of observations.

He had previously been active in many other ways. He had made three sea voyages, mainly to check upon what we now term magnetic variation; he had published papers on a wide variety of scientific subjects, and for a while he had even been Deputy Controller of the Mint at Chester. But, of course, it was his cometary work which ensured that he will never be forgotten. In 1706 he presented a paper to the Royal Society in which he claimed that the 1682 comet, which he had himself observed, was identical with comets previously seen in 1607 and in 1531, so that it had a period of 76 years. He predicted that it would return once more in 1758. This was a new idea; previously there had been no thought that some comets might be periodical. Halley could not hope to live long enough to see his prediction fulfilled; he died in 1742 – but the comet was recovered on Christmas Night, 1758, by the German amateur astronomer Palitzsch, and passed perihelion in 1759.

Halley had a cheerful, jovial personality, and his career was a story of almost unbroken success. Even apart from his cometary work, he would have earned an honoured place in the story of astronomy. So, as we await the coming return of 'his' comet, let us also pay due homage to Edmond Halley as a man.

PART THREE

Miscellaneous

Some Interesting Telescopic Variable Stars

Star	R.A. *h*	*m*	Dec. °	′	*Mag. range*	*Period, days*	*Remarks*
R. Andromedæ	0	22	+38	18	6.1–14.9	409	
W Andromedæ	2	14	+44	4	6.7–14.5	397	
Theta Apodis	14	00	−76	33	6.4–8.6	119	Semi-regular.
R Aquilæ	19	4	+ 8	9	5.7–12.0	300	
R Arietis	2	13	+24	50	7.5–13.7	189	
R Aræ	16	35	−56	54	5.9–6.9	4	Algol type.
R Aurigæ	5	13	+53	32	6.7–13.7	459	
R Boötis	14	35	+26	57	6.7–12.8	223	
Eta Carinæ	10	43	−59	25	−0.8–7.9	—	Unique erratic variable.
I Carinæ	09	43	−62	34	3.9–10.0	381	
R Cassiopeiæ	23	56	+51	6	5.5–13.0	431	
T Cassiopeiæ	0	20	+55	31	7.3–12.4	445	
X Centauri	11	46	−41	28	7.0–13.9	315	
T Centauri	13	38	−33	21	5.5–9.0	91	Semi-regular.
T Cephei	21	9	+68	17	5.4–11.0	390	
R Crucis	12	20	−61	21	6.9–8.0	5	Cepheid.
Omicron Ceti	2	17	− 3	12	2.0–10.1	331	Mira.
R Coronæ Borealis	15	46	+28	18	5.8–14.8	–	Irregular
W Coronæ Borealis	16	16	+37	55	7.8–14.3	238	
R Cygni	19	35	+50	5	6.5–14.2	426	
U Cygni	20	18	+47	44	6.7–11.4	465	
W Cygni	21	34	+45	9	5.0–7.6	131	
SS Cygni	21	41	+43	21	8.2–12.1	–	Irregular.
Chi Cygni	19	49	+32	47	3.3–14.2	407	Near Eta.
Beta Doradûs	05	33	−62	31	4.5–5.7	9	Cepheid.
R Draconis	16	32	+66	52	6.9–13.0	246	
R Geminorum	7	4	+22	47	6.0–14.0	370	
U Geminorum	7	52	+22	8	8.8–14.4	–	Irregular.
R Gruis	21	45	−47	09	7.4–14.9	333	
S Gruis	22	23	−48	41	6.0–15.0	410	
S Herculis	16	50	+15	2	7.0–13.8	307	
U Herculis	16	23	+19	0	7.0–13.4	406	
R Hydræ	13	27	−23	1	4.0–10.0	386	
R Leonis	9	45	+11	40	5.4–10.5	313	Near 18, 19.
X Leonis	9	48	+12	7	12.0–15.1	–	Irregular (U Gem type).
R Leporis	4	57	−14	53	5.9–10.5	432	'Crimson star.'
R Lyncis	6	57	+55	24	7.2–14.0	379	
W Lyræ	18	13	+36	39	7.9–13.0	196	

Star	R.A. *h*	*m*	Dec. °		*Mag. range*	*Period days*	*Remarks*
T Normæ	15	40	−54	50	6.2–13.4	293	
HR Delphini	20	40	+18	58	3.6– ?		Nova, 1967.
S Octantis	17	46	−85	48	7.4–14.0	259	
U Orionis	5	53	+20	10	5.3–12.6	372	
Kappa Pavonis	18	51	−67	18	4.0– 5.5	9	Cepheid.
R Pegasi	23	4	+10	16	7.1–13.8	378	
S Persei	2	19	+58	22	7.9–11.1	810	Semi-regular.
R Sculptoris	01	24	−32	48	5.8– 7.7	363	Semi-regular.
R Phœnicis	23	53	−50	05	7.5–14.4	268	
Zeta Phœnicis	01	06	−55	31	3.6– 4.1	1	Algol type.
R Pictoris	04	44	−49	20	6.7–10.0	171	Semi-regular.
L² Puppis	07	12	−44	33	2.6– 6.0	141	Semi-regular.
Z Puppis	07	30	−20	33	7.2–14.6	510	
T Pyxidis	09	02	−32	11	7.0–14.0	–	Recurrent nova (1920, 1944)
R Scuti	18	45	− 5	46	5.0– 8.4	144	
R Serpentis	15	48	+15	17	5.7–14.4	357	
SU Tauri	5	46	+19	3	9.2–16.0	–	Irregular (R CrB type).
R Ursæ Majoris	10	41	+69	2	6.7–13.4	302	
S Ursæ Majoris	12	42	+61	22	7.4–12.3	226	
T Ursæ Majoris	12	34	+59	46	6.6–13.4	257	
S Virginis	13	30	−6	56	6.3–13.2	380	
R Vulpeculæ	21	2	+23	38	8.1–12.6	137	

Note: Unless otherwise stated, all these variables are of the Mira type.

Some Interesting Double Stars

The pairs listed below are well-known objects, and all the primaries are easily visible with the naked eye, so that right ascension and declinations are not given. Most can be seen with a 3-inch refractor, and all with a 4-inch under good conditions, while quite a number can be separated with smaller telescopes, and a few (such as Alpha Capricorni) with the naked eye. Yet other pairs, such as Mizar-Alcor in Ursa Major and Theta Tauri in the Hyades, are regarded as too wide to be regarded as bona-fide doubles!

Name	*Magnitudes*	*Separation"*	*Position angle, deg.*	*Remarks*
Gamma Andromedæ	3.0, 5.0	9.8	060	Yellow, blue. B is again double (0".4) but needs a larger telescope.
Zeta Aquarii	4.4, 4.6	2.6	291	Becoming more difficult.
Gamma Arietis	4.2, 4.4	8	000	Very easy.
Theta Aurigæ	2.7, 7.2	3	330	Stiff test for 3 in. OG
Delta Boötis	3.2, 7.4	105	079	Fixed.
Epsilon Boötis	3.0, 6.3	2.8	340	Yellow, blue. Fine pair.
Kappa Boötis	5.1, 7.2	13	237	Easy.

SOME INTERESTING DOUBLE STARS

Name	*Magnitudes*	*Separation"*	*Position angle, deg.*	*Remarks*
Zeta Cancri	5.6, 6.1	5.6	082	
Iota Cancri	4.4, 6.5	31	307	Easy. Yellow, blue.
Alpha Canum Venat.	3.2, 5.7	20	228	Yellowish, bluish. Easy.
Alpha Capricorni	3.3, 4.2	376	291	Naked-eye pair. Alpha again double.
Eta Cassiopeiæ	3.7, 7.4	11	298	Creamy, bluish. Easy.
Beta Cephei	3.3, 8.0	14	250	
Delta Cephei	var, 7.5	41	192	Very easy.
Alpha Centauri	0.0, 1.7			Binary; period 80 years. Very easy.
Xi Cephei	4.7, 6.5	6	270	Reasonably easy.
Gamma Ceti	3.7, 6.2	3	300	Not too easy.
Alpha Circini	3.4, 8.8	15.8	235	PA, slowly decreasing.
Zeta Coronæ Borealis	4.0, 4.9	6.3	304	
Delta Corvi	3.0, 8.5	24	212	
Alpha Crucis	1.6, 2.1	4.7	114	Third star in low-power field.
Gamma Crucis	1.6, 6.7	111	212	Wide optical pair.
Beta Cygni	3.0, 5.3	35	055	Yellow, blue. Glorious.
61 Cygni	5.3, 5.9	25	150	
Gamma Delphini	4.0, 5.0	10	265	Yellow, greenish. Easy.
Nu Draconis	4.6, 4.6	62	312	Naked-eye pair.
Alpha Geminorum	2.0, 2.8	2	151	Castor. Becoming easier.
Delta Geminorum	3.2, 8.2	6.5	120	
Alpha Herculis	var, 6.1	4.5	110	Red, green.
Delta Herculis	3.0, 7.5	11	208	Optical double.
Zeta Herculis	3.0, 6.5	1.4	300	Fine, rapid binary.
Gamma Leonis	2.6, 3.8	4.3	121	Binary; period 400 years
Alpha Lyræ	0.0, 10.5	60	180	Vega. Optical; B faint.
Epsilon Lyræ	4.6, 6.3	3	005	Quadruple. Both pairs
	4.9, 5.2	2.3	111	separable in 3 in. OG
Zeta Lyræ	4.2, 5.5	44	150	Fixed. Easy double.
Beta Orionis	0.1, 6.7	9.5	205	Rigel. Can be split with 3 in.
Iota Orionis	3.2, 7.3	11	140	
Theta Orionis	6.0, 7.0			The famous Trapezium in M.42
	7.5, 8.0			
Sigma Orionis	4.0, 7.0	11.1	236	Quadruple. D is rather
		12.9	085	faint in small apertures.
Zeta Orionis	1.9, 5.0	3	160	
Eta Persei	4.0, 8.5	28.5	300	Yellow, bluish.
Beta Phoenicis	4.1, 4.1	1.3	352	Slow binary.
Beta Piscis Austrini	4.4, 7.9	30.4	172	Optical pair. Fixed.
Alpha Piscium	4.3, 5.3	1.9	291	
Kappa Puppis	4.5, 4.6	9.8	318	Again double.
Alpha Scorpii	0.9, 6.8	3	275	Antares, Red. green.
Nu Scorpii	4.2, 6.5	42	336	
Theta Serpentis	4.1, 4.1	23	103	Very easy.
Alpha Tauri	0.8, 11.2	130	032	Aldebaran. Wide, but B is very faint in small telescopes.
Beta Tucanæ	4.5, 4.5	27.1	170	Both components again double.
Zeta Ursæ Majoris	2.3, 4.2	14.5	150	Mizar. Very easy. Naked eye pair with Alcor.
Alpha Ursæ Minoris	2.0, 9.0	18.3	217	Polaris. Can be seen with 3 in.
Gamma Virginis	3.6, 3.7	4.8	305	Binary; period 180 yrs. Closing.
Theta Virginis	4.0, 9.0	7	340	Not too easy.
Gamma Volantis	3.9, 5.8	13.8	299	Very slow binary.

Some Interesting Nebulæ and Clusters

Object	R.A.		Dec.		*Remarks*
	h	*m*	°		
M.31 Andromedæ	00	40.7	+41	05	Great Galaxy, visible to naked eye.
H.VIII 78 Cassiopeiæ	00	41.3	+61	36	Fine cluster, between Gamma and Kappa Cassiopeiæ.
M.33 Trianguli	01	31.8	+30	28	Spiral. Difficult with small apertures.
H.VI 33–4 Persei	02	18.3	+56	59	Double cluster; Sword-handle.
△142 Doradûs	05	39.1	−69	09	Looped nebula round 30 Doradûs. Naked-eye. In Large Cloud of Magellan.
M.1 Tauri	05	32.3	+22	00	Crab Nebula, near Zeta Tauri.
M.42 Orionis	05	33.4	−05	24	Great Nebula. Contains the famous Trapezium, Theta Orionis.
M.35 Geminorum	06	06.5	+24	21	Open cluster near Eta Geminorum.
H.VII 2 Monocerotis	06	30.7	+04	53	Open cluster, just visible to naked eye.
M.41 Canis Majoris	06	45.5	−20	42	Open cluster, just visible to naked eye.
M.47 Puppis	07	34.3	−14	22	Mag. 5,2. Loose cluster.
H.IV 64 Puppis	07	39.6	−18	05	Bright planetary in rich neighbourhood.
M.46 Puppis	07	39.5	−14	42	Open cluster.
M.44 Cancri	08	38	+20	07	Præsepe. Open cluster near Delta Cancri. Visible to naked eye.
M.97 Ursæ Majoris	11	12.6	+55	13	Owl Nebula, diameter 3′. Planetary.
Kappa Crucis	12	50.7	−60	05	"Jewel Box"; open cluster, with stars of contrasting colours.
M.3 Can. Ven.	13	40.6	+28	34	Bright globular.
Omega Centauri	13	23.7	−47	03	Finest of all globulars. Easy with naked eye.
M.80 Scorpii	16	14.9	−22	53	Globular, between Antares and Beta Scorpionis.
M.4 Scorpii	16	21.5	−26	26	Open cluster close to Antares.
M.13 Herculis	16	40	+36	31	Globular. Just visible to naked eye.
M.92 Herculis	17	16.1	+43	11	Globular. Between Iota and Eta Herculis.
M.6 Scorpii	17	36.8	−32	11	Open cluster; naked-eye.
M.7 Scorpii	17	50.6	−34	48	Very bright open cluster; naked eye.
M.23 Sagittarii	17	54.8	−19	01	Open cluster nearly 50′ in diameter.
H.IV 37 Draconis	17	58.6	+66	38	Bright Planetary.
M.8 Sagittarii	18	01.4	−24	23	Lagoon Nebula. Gaseous. Just visible with naked eye.
NGC 6572 Ophiuchi	18	10.9	+06	50	Bright planetary, between Beta Ophiuchi and Zeta Aquilæ.
M.17 Sagittarii	18	18.8	−16	12	Omega Nebula. Gaseous. Large and bright.
M.11 Scuti	18	49.0	−06	19	Wild Duck. Bright open cluster.
M.57 Lyræ	18	52.6	+32	59	Ring Nebula. Brightest of planetaries.
M.27 Vulpeculæ	19	58.1	+22	37	Dumb-bell Nebula, near Gamma Sagittæ.
H.IV 1 Aquarii	21	02.1	−11	31	Bright planetary near Nu Aquarii.
M.15 Pegasi	21	28.3	+12	01	Bright globular, near Epsilon Pegasi.
M.39 Cygni	21	31.0	+48	17	Open cluster between Deneb and Alpha Lacertæ. Well seen with low powers.

Our Contributors

Professor Alec Boksenberg, Director of the Royal Greenwich Observatory, is one of the outstanding astronomers of modern times, and has been largely responsible for developing the IPCS, or Image Photon Counting System, which is of fundamental importance in modern research.

Dr Paul Murdin has made many contributions to astrophysics. He has carried out his researches at the Anglo-Australian Observatory and at Greenwich. He is now Project Scientist for the British telescopes at the new observatory on La Palma, Canary Islands.

Dr David A. Allen, one of our most regular and valued contributors, continues his research work at the Anglo-Australian Observatory in Australia. He is also the author of several popular books and papers as well as his technical work.

Dr Martin Cohen is concerned mainly with infrared astronomy, and is at present at the NASA Ames Research Centre at Moffett Field, California.

Dr Steven W. Squyres and **Dr Ray T. Reynolds** are researchers at the NASA Ames Research Centre in Mountain View, California. The work described in their article was done in conjunction with Patrick Cassen, David Colburn, and Christopher McKay at Ames, and Stanton Peale at the University of California, Santa Barbara.

Professor Clyde W. Tombaugh will always be remembered for his discovery of the planet Pluto, from the Lowell Observatory at Flagstaff (Arizona) in 1930, but this has been only one of his many contributions to astronomy. He is now Professor Emeritus at the University of New Mexico, Las Cruces.

F. G. Watson, of the Royal Observatory Edinburgh, is a specialist in astrophysical research. He is at present working at the UK Schmidt Telescope in Australia.

Steven A. Bell is a member of the Department of Astronomy at the University of St Andrews in Scotland.

Patrick Moore is President of the British Astronomical Association for the 1982–1984 sessions.